TOTAL CUSTOMER RELATIONSHIP MANAGEMENT

BY
MITCH SCHNEIDER

THOMSON
DELMAR LEARNING

Australia • Canada • Mexico • Singapore • Spain • United Kingdom • United States

Automotive Service Management

Total Customer Relationship Management

Mitch Schneider

Executive Director:
Alar Elken

Executive Editor:
Sandy Clark

Automotive Product Development Manager:
Timothy Waters

Development:
Kristen Shenfield

Executive Marketing Manager:
Maura Theriault

Channel Manager:
Beth A. Lutz

Marketing Coordinator:
Brian McGrath

Executive Production Manager:
Mary Ellen Black

Production Manager:
Larry Main

Production Editor:
Elizabeth Hough

Editorial Assistant:
Kristen Shenfield

Cover Design:
Julie Lynn Moscheo

Printed in Canada
3 4 5 XX 06 05 04 03

For more information contact Delmar Learning
Executive Woods
5 Maxwell Drive, PO Box 8007,
Clifton Park, NY 12065-8007
Or find us on the World Wide Web at
http://www.delmar.com

Library of Congress Cataloging-in-Publication Data:

Schneider, Mitch.
Automotive service management. Total customer relationship management / Mitch Schneider.
p. cm.
ISBN 1-40182-657-1
1. Service stations--Management. 2. Customer relations. I. Title: Total customer relationship management. II. Title.
TL153 .S3624 2003
629.28'6'068--dc21

2002151915

NOTICE TO THE READER

Publisher does not warrant or guarantee any of the products described herein or perform any independent analysis in connection with any of the product information contained herein. Publisher does not assume, and expressly disclaims, any obligation to obtain and include information other than that provided to it by the manufacturer.

The reader is expressly warned to consider and adopt all safety precautions that might be indicated by the activities herein and to avoid all potential hazards. By following the instructions contained herein, the reader willingly assumes all risks in connection with such instructions.

The publisher makes no representation or warranties of any kind, including but not limited to, the warranties of fitness for particular purpose or merchantability, nor are any such representations implied with respect to the material set forth herein, and the publisher takes no responsibility with respect to such material. The publisher shall not be liable for any special, consequential, or exemplary damages resulting, in whole or part, from the readers' use of, or reliance upon, this material.

CONTENTS

PREFACE

This shop operations guide is designed to provide a framework that helps you make consistent, high-quality, and productive service a part of your normal, everyday shop operations.

Is it everything you ever wanted to know about automotive shop management but were afraid to ask?

I'm not sure. In reality, it is a lot of what I have learned about running a successful automotive service business, both on my own and sitting at the feet of those who started this journey before I did. At the very least, it's everything you can fit into eight volumes.

The purpose of this guide is to help ensure that *great performance* is achieved every time you or one of your employees approaches a regular or potential customer to deliver automotive service. It is the only way I know to create trust and insure customer loyalty. And, great performance coupled with increased customer loyalty, trust, and operational excellence will almost always result in increased profits.

High-quality service is defined by the customer's experience while in your care and custody. Operational excellence ensures that the customer's cars will be *ready when promised* and *fixed right the first time*. Fixing the car right the first time is as much about understanding what your customer is trying to tell you when they are communicating their frustrations with the vehicle as it is about proper diagnosis and professional quality work. *Total Quality Service* is only possible through high-quality communication and knowing where you are in the process of completing a job at all times. Because the workforce is diverse, we chose to alternate between "he" and "she" from chapter to chapter to ensure gender is equally represented.

Selling service is perhaps the most difficult of all sales to make. Buying a product—something you can hold in your hand like an oil filter or a spark plug—constitutes the purchase of something you can touch or feel. Purchasing a service is something else entirely. When a vehicle owner buys a service, that person is really purchasing a promise—a promise that will be fulfilled in the future. That requires trust and a great deal of faith.

Your ability to perform is based upon a lot of things, not the least of which is technical competency. But you can't demonstrate that technical competency unless (or until) you are given the opportunity. The vehicle owner can't see technical competency. She can't touch it or feel it and you can't hang it out in front of the building like a sign.

The customer's perception of whether or not you are worthy of her trust, and whether or not you will be able to solve her automotive service problem, will be based in many cases upon any one of a number of subtle, almost subliminal, messages. These messages are broadcast and received before any one of your technicians ever picks up a screwdriver, wrench, ratchet, or socket.

Service starts when a customer sees and responds to an advertisement (an invitation to experience the quality of your service), hears about your business from a friend or relative, calls, or walks in, service includes every subsequent customer contact until the job is completed—ready when promised. Moments of truth—moments when a customer has the opportunity to come in contact with any member of your organization and form an opinion about you or your company—occur before, during, and after the job has been completed. They will ultimately define your personal success and the success of your company.

Consistent execution of company policies and procedures is the catalyst for world-class service. It is perhaps more important to be consistent at what you do than to be good at what you do. Do a great job one time, and something less than great the next, and you aren't likely to see that customer again. The more people you have working for you, the greater the challenge it is to achieve consistency. Without a comprehensive and complete set of policies and procedures, consistency is nothing more than an elusive dream.

All of this consistent high-quality service is only sustainable if your shop is operating at productivity levels that allow you to make a profit. Productivity is a by-product of your technicians' ability to execute policies and procedures flawlessly. It will take more than a visit to the tool truck or a one-day management seminar for your technicians to bill 70 percent or more of their available hours.

ABOUT THE SERIES

All eight volumes of this guide—*Total Customer Relationship Management, From Intent to Implementation, Managing Dollars With Sense, Operational Excellence, Building a Team, The High Performance Shop, Safety Communications, and Operations Management*—are designed to work together in large shops and small shops. That means that the building blocks of consistency—the policies and procedures you will find within these pages—are defined in their simplest terms because no matter how big or how small a shop is, the same kinds of tasks must be performed. The policies and procedures we define here can and will be carried out by the service writer, service manager, technician, office manager, and lot person. In many cases, the person who fills the role of service writer also serves as the service manager and perhaps even the technician. The key to success is not who does the job. The key to success is insuring that each critical task is completed accurately and consistently—the same way every time.

While this set of guides will take you through a series of step-by-step procedures, it is important to remember that delivering high-quality service is a process. To achieve success, all of the steps in the process must be performed, either consciously or unconsciously. If you are having a problem in one particular area of your business, start with the guide that addresses that need first. Next, read through the rest of the guides. See if what you find in these pages makes sense to you. *Then, read it again.* Look at your business and see what, if anything, applies to you. *Then, read it again*. Consider which of your current procedures need to remain in place. If some need to be changed, determine if some of the policies or procedures you found in these guides can be incorporated into the daily operation of your business. *Then, read it again.* Put yourself in the place of your customer. Involve everyone in your business in this exercise. You will be far more successful with their help and cooperation than you could ever be without it. Help them understand that this is how things need to be done from *this moment forward* and why. Walk through the new procedures in your shop. Pull the trigger, start the process, read the text, and begin your journey to a better quality of service, a stronger, healthier, more profitable business, and a more satisfying, less frustrating life.

ACKNOWLEDGEMENTS

Nothing of consequence occurs in a vacuum. It takes a lifetime of collaboration, cooperation, and support. The Automotive Service Management series is no exception. If it takes a village to raise a child, it takes a city to conceive, write, edit, and publish one book—let alone eight. Consequently, there are some special people I would like to thank for making the journey with me.

I'd like to thank my Delmar family: Tim Waters, Kristen Shenfield, Betsy Hough, and everyone in marketing for staying with me from conception to completion—as well as Lynanne Fowle and Megan Iverson at TIPS Technical Publishing for their meticulous attention to every detail. It is a joy to work with so many people who care.

I'd like to thank John Wick and Willi Alexander for planting the seeds, and Dick Vinet, the Duffy-Vinet Institute, Motor Service Magazine, and Champion Spark Plugs, for presenting the first shop management seminar I ever attended. It gave me confidence, but more than that—it gave me hope. And, Arthur Epstein (of blessed memory) and Ray Walker for their guidance, friendship, and patience and for recognizing in me something I could not or would not see myself.

Most of all I'd like to thank my wife, Lesley, for tolerating a lifetime of ruined dinners, working too late at the shop, and hiding out in the "Cave" writing and thinking and just generally being far away. She's not only been the biggest part of everything I've ever accomplished, believing even when I didn't—she's been the best part of everything I've ever accomplished. I'd like to thank my kids, Ryan and Dana, for sharing me with my work even when I'm sure they didn't understand. They are a blessing and an inspiration.

I'd like to thank my brothers: David for working with me for the better part of thirty-six years, and Todd for not. Both have taught me a great deal. David has taught me about patience and understanding and looking beyond my frustration to try to understand the lesson he is trying to teach me about myself. And, Todd has taught me a great deal about positive attitude, professionalism and personal courage, in addition to never saying No to anything I've ever asked of him.

Finally, I would like to acknowledge my parents—Sylvia and Jerry Schneider. They are far more than parents. They are my closest friends, my favorite teachers, and a living example of what excellence, character, commitment, dedication, sacrifice, and giving everything you have to give is all about. It's hard to relax and take it easy when your eighty-year-old parents won't, and it's just as hard to give less than your best when your eighty-year-old parents won't!

INTRODUCTION

Survival in a complicated and changing world is all about assumptions. Assumptions allow us to function when we find ourselves beyond the limits of our understanding and experience. When we talk about *paradigms*, we're just defining these assumptions in more scientific language. Our assumptions—our paradigms—are not based upon the world as it is. They are based upon the world as we perceive it to be based upon our experience, the information available, and our understanding of that information at that moment in time. Change the information and you change the world—or at least our perception of the world.

Success in business can be all about assumptions as well. We surround ourselves with a wall of assumptions to help us make sense of the chaos we confront every day. To a large degree, our success is based upon just how accurate many of those assumptions are. One of the ways we do that is to provide service based upon what we believe is best for the vehicle, assuming that what is best for the vehicle is best for the consumer when that might not necessarily be the case.

In the end, these assumptions are about what we *believe* the customer wants, needs, and expects from us, not necessarily what the customer's *actual* desires and expectations are.

A perfect example of this is the constant battle between *ready when promised* and *fixed right the first time*. These two often jockey for position as the *number one customer concern* when vehicle owners are seeking automotive service. You and I want the vehicle to run as well as it possibly can. We believe that the customer wants the same thing and, generally, that's true. But, more than that, the customer really wants the vehicle delivered when we said it would be delivered. Life is complicated. Time is precious. The logistics of having a car in the shop can devastate the careful planning most of us have built into our lives.

Where many of us might choose *fixed right the first time* as the number one customer concern, research suggests that our customers would more likely choose *ready when promised*. And, where exactly, does that leave all the shop owners and managers who insist that *price* is the customer's first and only consideration?

Total Customer Relationship Management is all about the relationship that exists between the provider of automotive products and services and the recipient of those products and services. It is all about questioning our assumptions and then re-creating a service environment that is responsive and respectful when it comes to those things that are most important to our customers and clients. It's about who we are and who we will need to become just as much as it is about customer expectations, retention, loyalty, and satisfaction.

In the end, it is about creating a compelling value equation—something your target customer will not be able to resist—and then delivering your services and products in a total quality service environment. It is about managing the whole scope of your relationship with the vehicle owner, recognizing that perception is reality and that the only perception that matters is the perception of your customers.

So, take this as an opportunity to look at your relationship with the vehicle owner in a new and different way. Confront your assumptions about customers, business, and our industry and consider how different things could be if we just change the way we look at them.

Remember, if you change the way you see things—if you change your paradigms and your assumptions about your customers, your business, and our industry—you just might change everything... and in an industry like ours where confidence, self-image, and profits have been notoriously low, that might not be such a bad thing!

A Little Bit About Customer Relations

INTRODUCTION

This chapter introduces you to:

- The history behind customer relations
- One-to-one marketing

WHAT IS CUSTOMER RELATIONS?

Who are these people and what do they want from us? What is the *right* kind of customer and where do we find them? What do we do with them if we are fortunate enough or persistent enough to develop the skills necessary to bring them to our door? Why are we so preoccupied with the wants, needs, and expectations of our clients? Aren't there plenty of customers and potential customers out there?

If I own an automotive repair shop, isn't the primary purpose of my business to service, maintain, and repair cars and trucks? Isn't doing that well—doing it better than anyone else—enough? If it isn't, shouldn't it be? And, as long as we're on the subject, are we really talking about customer *relations* (what we do in relation to what the customer wants, needs, or expects from us), or are we talking about customer *relationships* (the relationship we hope to build with the customer), which is and should be virtually all-inclusive? Why focus on customers anyway? Are there other things just, or at least, as important?

Peter Drucker, guru of American business management writers, has said that the only purpose of business is to attract customers and to innovate (products and services)—anything not specific to making customers and innovating becomes a *cost* of business, not its purpose. Proctor & Gamble, perhaps one of the premier marketing organizations in the world, suggests that the purpose of marketing (everything involved in bringing a potential customer to your door) is to provide products and services that solve people's problems at a profit. Theodore Levitt, a well-known professor of business management and former editor of the Harvard Business Review, says the difference between marketing and sales is pretty simple—sales is the process of getting rid of stuff you don't want, while marketing is the process of letting people know you have the stuff they both want and need.

All three of these concepts and every one of the preceding questions have something very basic in common. They all involve someone other than yourself—the customer—who responds to your innovative products and services, buys them, makes your business life possible and, as a consequence, most of the rest of your life as well.

For years, these customers were taken for granted, ignored, or considered *part of the territory*—a necessary evil. They were, after all, curious, demanding, and sometimes annoying. They called incessantly, came by unexpectedly, and questioned us endlessly. In general, they were considered a pain—a *cost* of doing business. Of course, that was until they started making other choices, taking their business elsewhere to individuals or companies more responsive to their wishes, needs, and expectations. With their absence, the value they represented became instantly apparent. And, at that point a new movement began. It was a shift in business philosophy—a move away from mass marketing toward one-to-one marketing.

I believe that, in order to understand our new and changing relationship with the consumer, we must first understand this paradigm shift in American business and business in general. We must define this relationship with the consumer if we ever hope to manage it. If we are successful, a whole new world of customer interaction is possible. If we are not, we are doomed to a world of shrinking margins, inadequate profits, and lost opportunities.

When I started thinking and writing about our relationship with the consumer almost twenty years ago, very little material was available on automotive shop management and even less on our relationship with the vehicle owner. Interestingly enough, this field still does not have much substance almost twenty years later, which was (and perhaps still is) the major catalyst for undertaking this project in the first place.

For us to put the whole universe of automotive shop management into some kind of perspective, we have to begin with a new and different understanding of who we are, who our customers are, how our relationship with those customers has changed, and how it all happened without the majority of us seeing it. That is why I think it is a good idea to take a quick look at the past before we begin our exploration of either the present or the future.

The Olde Days of Automotive Service Management

At the end of the Agrarian Age, the stars aligned to bring cheap labor, developing technology, and raw materials together in Europe and North America. The result was the beginning of a new age—the Industrial Age. With an abundance of raw materials, people, and facilities, all that was needed were buyers—*people* to buy the new products and services made possible by expanding manufacturing capacity. I stress *people* because, as it is used here, it is a fairly impersonal term.

Buyers were all that was necessary to fuel the transition from the farm to the factory. Relationships between consumers and producers were uncomplicated and almost unnecessary. Intimate relationships with those buying your products and services weren't altogether necessary for most manufacturers. Because of the high demand, all they had to do to move products was to make those products and services available.

The marketplace continued to change—moving from individuals, caring, and craftsmanship to mass production, convenience, and availability. The corner grocery store was replaced by the supermarket, the local men's shop was replaced by a department store or warehouse, and the corner pharmacy was replaced by the drug store as more and more products found their way into fewer, yet larger, locations.

Even as demand began to level off, all manufacturers had to do to maintain a steady flow of products and services was innovate and/or create a preference based upon the product's brand—some reason other than, or in addition to, utility. Even with the development of *branding*, the process was basically transactional in nature, requiring little or no intimacy. All you needed was desire (whether manufactured and contrived, or genuine), product availability, and a consumer with the opportunity and necessary resources, and the goods and services continued to flow.

One-to-One Marketing

As time passed, competitive advantages became more and more difficult to identify, let alone sustain. A new differentiator emerged—a differentiator that had, interestingly enough, been there all along. That differentiator was, and still is… relationships—relationships between manufacturer and wholesaler, wholesaler and retailer, and, most importantly, retailer and consumer. The retailer-consumer relationship is what one-to-one marketing is all about. For those of us in the service industry, the best part of this new paradigm is that this is where we really shine because we have been doing one-to-one marketing since the dawn of time.

You see, in the Industrial Age, mass marketing is the key to moving the greatest amount of product to and through the marketplace. It's all about having the greatest percentage of the total universe of potential customers. It's all about creating brand awareness and then creating a fury of demand for your products and services. With mass marketing, who buys those products and who takes advantage of those services is almost inconsequential, except for those doing demographic research and then only in regard to branding.

Mass marketing is all about the numbers—nothing else matters. It is all about *pushing* products through distribution channels. It is an almost perfect match for transactional relationships, which aren't really relationships at all. In a transactional business environment, making the sale and moving the merchandise is the only objective.

One-to-one marketing is the polar opposite. It's about as far away as you can get from mass marketing. In our case, it asks a simple question: What percentage of all the automotive service dollars spent by an individual or family are you currently getting? And, if you are not getting a high percentage of those sales, why aren't you?

If making the sale, moving the product, and forcing the outcome have the potential to damage the relationship with a customer, a good one-to-one marketer will respectfully back away and might even recommend a competitor better suited to satisfy the customer. The customer's wants, needs, and expectations, as well as your need for a long and profitable relationship with that customer, should be at the core of every action and decision. Everything else comes second.

This new interest in the relationship that exists between businesses and their respective customers has created a new industry—customer relationship management. At its very heart is the use of customer data to maximize the profitability of each customer relationship. It suggests that through the use of this data, you can better serve the customer by better understanding who they are and what they want. I believe with a perfect faith that our industry's future will ultimately rely on our ability to embrace this new doctrine. However, having said that, I also believe that it is critical to understand that the only things you and I can manage in our relationship with the consumer are ourselves, our companies, and the ways we interact with the customer before, during, and after the interchange takes place.

CHAPTER SUMMARY

We certainly can't manage the customer and how he behaves. If we do, we risk everything. No one likes to be manipulated. No one likes to be led or forced into taking any action. Consequently, we must be where the consumer wants to go—with the products and services he wants to buy, at prices he is willing to pay—and deliver a quality of service that makes him want to come back. That is what customer relations and customer relations management are all about.

Our task here is to explore our relationship with the vehicle owner and how we can insure our success by combining world class products and skill with world class service.

How will you successfully manage your relationship with the customer? Write your thoughts in the chart on the following page:

Thoughts

Actions

Results

CHAPTER 2

A Vision For The Future

INTRODUCTION

This chapter introduces you to:

- The importance of a mission statement and how it reflects the values and principles of the company it represents
- Vision statements, objectives, and guiding principles

MISSION STATEMENT

Once upon a time, a very long time ago, I heard something that really changed my life. It was a simple statement of fact that suggested that

"If you always do what you always did, you will always get what you always got!"

At first, it really didn't register. In fact, I remember thinking that it was a little silly. But the more I thought about this simple little phrase, the more I realized just how profound it is. You see, until then my knowledge was, for the most part, theoretical. I read a lot and even thought about the different ways I could apply what I learned to a business, but I never took the next logical step to do anything about it. I just waited. I waited for someone else to do it. I waited for someone else to come and fix my business. I waited for someone else to come and fix my industry. And I waited.

Then, all of a sudden, I realized that it was *my* responsibility to act, and that my paralysis and my lack of action might be the reason nothing good was happening. It might be the reason nothing was happening at all. In fact, it all became very clear, very quickly. I suddenly knew that it was my responsibility to make a difference, and no one else's.

"If you always do what you always did....."

You have to admit, it was a strange call to action. The only questions remaining were what to do differently and how. Which direction do you face when you realize it's finally time to start your journey? The answer begins with another question: Where do you want to go? To start the journey, you must have a destination, and that destination will determine your direction and what you need to make the journey. I think that's where vision comes into play. Your vision becomes the destination... your destination... the target. It is the *where* and the *why* of making the journey in the first place. It is what you see in your mind's eye when you think about your business in the context of the future—a different future. And, I think it is important enough to discuss here at the beginning of this guide.

It is the *mission* of every company to achieve its *vision*, just as it is the mission of every person to achieve her personal vision for her own future and for the future of her family. This passion and commitment should be captured in a simple statement that clearly defines the overriding purpose of the company (or the person). Mission statements appear on the walls and in the hallways of most major corporations around the world. It is impossible to escape them. They seem obvious, espousing values that no one could reasonably refute. You can't go to a bank, a department store customer service window, or a hotel without being forced to confront the company vision or mission statement pointing out just how far from reality actual performance can be. The problem is that most mission statements do not have anything to do with reality. They are filled with lofty platitudes, all right, but they generally do not reflect the way upper management or production workers view what they are doing—especially in relation to how it impacts the customer or the culture in which they function. A mission statement should reflect the most basic **values** and principles of the company—who you are and what you strive to be. It should

val·ue (val'yoo) *noun Abbr.* val. 4. A principle, standard, or quality considered worthwhile or desirable: *"The speech was a summons back to the patrician values of restraint and responsibility." (Jonathan Alter)*

encompass four basic areas of human needs: *economic* or monetary-based needs, *social* or relationship-based needs, *psychological* or personal growth-related needs, and *contribution* or spiritual needs.

The Schneider Automotive mission statement developed over time and appeared on a number of our promotional pieces and advertisements before we ever realized it was a mission statement at all. It stated what we wanted to accomplish clearly enough, but we still never thought of it as a mission statement until we realized just how hard it would be to come up with something that reflected what we were trying to accomplish any more clearly. Our mission is...

> *To become as important to our clients as they are to us by providing them with the needed work, done well, finished on time, and at a fair price, with all work guaranteed.*

Our mission statement does not encompass all four of the needs mentioned above, but we believe our vision statement—our vision of the future as we would like it to be—does.

Vision

Volumes have been written about *vision*, and yet very few can define it clearly enough for it to be shared. At its simplest, vision is what *might* be, what *could* be, and where we *would like to be* at some point in the future. It is a statement of the future that could, should, or might be. It can exist in the mind of the individual or it can be written down. It can be intimately personal or it can encompass the vision of a business, an industry, a nation, or even the world we share.

For our purposes, we will limit our discussion to the vision that begins with an individual and then is applied to a business like yours or mine. Whether we like to admit it or not, all of us have an idea of what the future could look like *if everything went right for a change*. It might mean a bigger house, a better car, a new set of golf clubs, or the time to play. It might include loved ones... or it could be selfish. Regardless, all of us have a tendency to project ourselves into the future and all of us possess the imagination to bend that future (at least in our minds) until it makes us smile.

Creating a vision begins with a question that each of us asks before we embark upon any journey: Where do we want to go?

One of my favorite passages in literature addresses that issue at its very core. Unfortunately, it is appropriate for the majority of people we come in contact with on a daily basis. It comes from Lewis Carroll's *Through the Looking Glass*. When Alice realizes just how lost she has become, she meets the Cheshire Cat and asks for help. Their exchange goes something like this...

> *The Cat only grinned when it saw Alice. It looked good-natured, she thought: still it had VERY long claws and a great many teeth, so she felt that it ought to be treated with respect.*
>
> *"Cheshire Puss," she began, rather timidly, as she did not at all know whether it would like the name: however, it only grinned a little wider. "Come, it's pleased so far," thought Alice, and she went on. "Would you tell me, please, which way I ought to go from here?"*
>
> *"That depends a good deal on where you want to get to," said the Cat.*
>
> *"I don't much care where—" said Alice.*
>
> *"Then it doesn't matter which way you go," said the Cat.*
>
> *"—so long as I get SOMEWHERE," Alice added as an explanation.*
>
> *"Oh, you're sure to do that," said the Cat, "if you only walk long enough."*

If you don't know where you are going, any road will do!

For most people, that seems to be good enough and all too true. A clear vision of the future—your future or the future of your business—eliminates the confusion. It gives you a destination and your subconscious mind the opportunity to chart a course to it. Simply verbalizing that vision—stating it publicly—can be the magical beginning of an unbelievable adventure.

Where do you want to be in a year or two? In five? In fifteen? What will your business look like?

What will it feel like when you are there—to realize what you have built? What will it be like to work there? And, perhaps most importantly, what will it be like to do business with your company from the point of view of your customers? If you can close your eyes and see a picture of that business, you are light years ahead of your competitors. But first you have to think about that business in a way you have never thought about it before. You have to see it in the future, not in the present. That means you might see that business in a new location, or maybe even in a different city or state, but you still have to be able to see it.

My vision was simple. This is how I saw our company in the future...

Vision Statement

To reinvent Schneider's Automotive Repair into a total quality service-oriented, customer-driven company, focused on providing quality automotive service to our customers, improved quality of life for our employees, and increased profits to our stakeholders by doing what we do best:

Providing the needed work, done well, finished on time and at a fair price, with all work guaranteed

There is one other thing... a small but important caveat. There must be a good *why* to produce the effort required to fuel change and, subsequently fuel growth. Both come with a price and that price is a certain level of discomfort, sacrifice, and more than a little hard work. The *why* can provide the strength required to continue when quitting would seem a much more reasonable alternative.

Generally speaking, the *why* should be very personal in nature—a clear picture of the rewards that come from sacrifice and success: a bigger house, a better car, or more time with family and friends. Without an adequate *why*, the sacrifice can become unbearable and the effort exhausting.

Goals

goal (gōl) *noun* 1. The purpose toward which an endeavor is directed; an objective. See synonyms for intention.

Goals are the major steps you take in order to achieve your vision. They can be either short- or long-term in nature, but they must be written down and published publicly (unless they are personal goals). By published publicly, I mean hung on an office wall or shared with someone whose opinion you regard highly. Here is what just a few of our goals looked like at Schneider's Automotive in early 1996:

To reinforce our position as the market leader in each of the following areas of high-quality automotive service in our community:

- High-tech diagnosis and repair
- Wheel service and undercar
- Air conditioning and comfort
- Emissions service and repair

To increase gross sales by a minimum of 10% per year for the next 4 years.

We had managed to achieve a number of these goals by the time I sat down to write this guide. Establishing these goals and sharing them among the principals in the company was one of the reasons for our success. Another was transmitting these goals to our employees in the way of expectations. Still another was ensuring that everyone understood the rewards for success and the consequences of failure. And finally, ensuring that these goals were realistic and attainable also helped us achieve success.

ob·jec·tive (ob-jek-tiv) *adjective, noun* **1.** Something that actually exists. **2.** Something worked toward or striven for; a goal. See synonyms for intention.

Objectives

Objectives are the small steps you take to bring you closer to your goals. It is wonderful to have a goal of increasing gross sales by a minimum of 10 percent per year for the next four years. It's even better to write the goal down and share it with everyone necessary to make the goal a reality. But, it still doesn't answer the question of how you are going to do it. Achieving your goal takes planning and the institution of smaller steps that will bring you incrementally closer to your goal every day.

At Schneider's Automotive, we felt we could achieve that kind of growth if we:

- Increased car count by 2 cars per day
- Increased gross profits by 10%
- Increased the number of vehicle inspections to a minimum of 3-5 per week (more about vehicle inspections later)
- Decreased or recovered expenses by 1%

There are libraries filled with books on how to set goals and objectives and still more on creating a vision or mission statement. You can make it as easy or as difficult as you want it to be. You can take all of your key people and lock yourselves in a room until you have hacked out a mission and/or vision statement that all of you can embrace and align yourselves with. Or, you can borrow mine until you come up with a better one. Whatever you do, please make sure that it reflects what you want to *be* in the future. Make sure they are consistent with your shared values and that they are reflected in the way you treat your employees and your customers. There can be no gap between vision and systems.

That brings us to another component of successful business operations—values. Values are the moral and ethical compass by which a business operates and by which its every decision can be judged or measured.

Values and goals are not the same thing. Goals are something you hope to achieve, while values are something you believe in. Goals can change. Values, like courage, character, integrity, honor, responsibility, and service, should not.

Guiding Principles

When values are institutionalized by a company, they become the principles that guide it, and these *guiding principles* will determine just what kind of a company it will be—how it will treat its customers, its employees, its vendors and suppliers, and how it will interact with the community it serves.

The following guiding principles (found in the book *It's Not The BIG That Eat The SMALL, It's The FAST That Eat The SLOW!* by Jason Jennings and Laurence Haughton) belong to Charles Schwab & Co.

1. Is it fair to our customers?
2. Does it respect our fellow employees and the spirit of teamwork?
3. Are we striving relentlessly to improve what we do and how we do it?
4. Will it earn, and will it be worthy of, our customers' trust?
5. Will it reinvent the business?
6. Are we willing to risk short-term revenue to do the right thing for the customer and ensure long-term success?
7. Will we own the technology?
8. Does it leverage the brand to build trust?
9. Will it create and nurture a spirit of innovation?

The following are the guiding principles of Lend Lease Corporation, and are also found in Jennings & Haughton's book.

1. Dare to be different in everything we do—become the enemy of mediocrity!
2. Never do anything that would diminish the pride our parents have in us!
3. No nasty surprises—grow earnings every year!
4. Be a leading employer.
5. Enhance the environment.
6. We need special relationships to enhance our capabilities.
7. No individual has a monopoly on good ideas.
8. We will only prosper with the support of the communities with which we interact.
9. We would like all employees to be shareholders.
10. We believe there is a strong link between good governance and performance.

These are the principles that, when applied, guide every decision made by anyone in the organization.

Where do they come from? They should come from you! They should flow from the way you see your world—how you would like to be treated and how you insist on treating others.

They are a statement of how you see the world around you—good and bad, right and wrong, moral and immoral. It defines the role of your company in the community it serves, and it recognizes the importance of all those around you: customers, employees, vendors, and suppliers.

How are guiding principles enforced? Before they can be enforced, they must be observed and experienced. Leaders demonstrate by example—they do not demand. When everyone in the organization internalizes the guiding principles of the leader (the president, chief executive officer, or the individual responsible for driving the company forward), these principles become an integral part of the company. They are not just words that are spoken—they are lived out every day.

The process begins by doing as you say and by living your principles and values—not just talking about them. It begins with consistency and, generally, ends with respect!

It begins with you. It's *YOUR* Job... it's *YOUR* company! *So...*

1. Start the process by creating and then publishing a set of guiding principles and then live by them!
2. Try to make sure that all decisions are consistent with those guiding principles.
3. Make sure every one knows and understands what the guiding principles are and the consequences of acting against them!
4. Make sure everyone understands that they must all act in accordance with them at all times.
5. Discuss them frequently.
6. Never publicly reward or acknowledge anyone who acts against them, even if those actions result in short-term gain.

CHAPTER SUMMARY

What is your vision for the future? What is your mission? What goals do you hope to achieve? What are your objectives? What principles guide you and everyone in your organization?

Take a moment to think about a different future and what it would be like to live in that future, and then write your thoughts in the chart on the next page.

Thoughts

Actions

Results

CHAPTER 3

Understanding Our Relationship With The Consumer

INTRODUCTION

This chapter introduces you to:

- Our relationship to vehicle owners and what we offer them from a business perspective
- The five pillars of trust: predictability, accountability, reliability, responsibility, and integrity

WHAT DO YOU AND I SELL?

What really is the basis of our relationship with the vehicle owner? I ask this question every time I do a Service Dealer Automotive Shop Management seminar, and the answers I get would amaze you. Before we talk about each one of them in depth, I'd like to mention that while all of the answers are correct in a way, they are all equally just as wrong. It's safe to say that barely ten percent of the garage owners or operators I have met truly understand the real nature of what the customer is actively seeking to purchase and what he wants out of this dynamic and complicated relationship he shares with us. Consequently, most of our efforts are, at best, misdirected and consequently ineffective.

Time

When I ask what we offer the motoring public each time we open our doors for business, *time* is one of the most common answers. There are few professions more neurotic about the sale of time than we are. We are so fixated with time that we are not content to sell it by the hour or quarter-hour—we have to break it up into six-minute tenths. Yet, as neurotic as we are about the sale of time, few if any of us ever keep track of how much time is actually spent working on a vehicle—or, for that matter, how much time is lost. Nor, oddly enough, is the customer particularly interested in how much time he or she is being charged for... unless there is a problem.

Most people only care whether or not the vehicle can be delivered when promised and whether or not the problem is solved when the vehicle is returned. In our experience, they are certainly more interested in the total cost of the repair than they are in the individual cost of each component of the repair.

Specialized Tools and Equipment

The use of specialized tools and equipment is another common answer to the question of what we offer our customers. We *do* use specialized tools and equipment every day—it is almost impossible to work on today's cars and trucks without using high-tech equipment. And yet, all too often garage owners across the country refuse to figure the cost of that equipment into their hourly labor charge, nor are they willing to charge separately for the use of that equipment. It is purchased and it is used, but all too often the Return On Investment (ROI) that is necessary to sustain a business and invest in future high-tech equipment purchases is lost forever.

High-Touch/High-Tech

Do we sell high-touch/high-tech? Is that what the customer wants? Well, high-tech is easy enough to understand—it is the constant and relentless introduction of space-age electronics to every system on the vehicle. Even today's *economy* cars are likely to have more than one computer, distributor-less or direct ignition, electronically controlled transmissions, active suspensions, fuel injection, and more. We sell high-tech because we live with it every day. It is a part of everything we do. We also sell high-touch, which is the time and skill it takes to explain all these changes to a frustrated and unaware motoring public.

High-touch/high-tech is a big part of what we do, but it is not what the motoring public wants or really needs when they come to us for service, maintenance, or repair.

Physical Plant

We cannot do the work unless we have some place to do it! Because of this, *physical plant* plays a very important role in everything we do. The facility has to meet or exceed customer expectations. In today's world—a world in which 65 percent of all the automotive service work is brought into the shop environment by a woman—meeting expectations means a clean and uncluttered environment and an inviting, efficient, and professional-looking shop. Nothing less will do, as the bar for quality-of-service delivery is raised throughout our society in every industry, not just ours. Do we think about what it is going to cost to paint the building next year? Has anyone considered what it will cost to put a roof on the building or new light fixtures in the shop? Are any of these costs reflected in the cost of the services we provide the customer? In almost every case, the answer is No.

People don't come back to us because of the building or the paint job or the roof, but they may not give us that necessary *first chance* because of the way the shop looks—especially if it looks unprofessional, disorganized, and chaotic.

Competence

How does competence affect service? We would all like to think everyone in our profession is competent, but are they? We all know the answer to that question is No. And yet, the customer believes that regardless of whom they have chosen for service, that person is competent or will supply competent craftsman to work on their vehicle. Competence in our industry is almost assumed, despite the bad press we receive on a regular basis. The customer does not go out of his way to find it because he believes that he is entitled to it—it is a given… an assumption.

Skill

Skill is almost taken for granted in our industry, which is really odd because it is so rare. More to the point, how the skill is acquired and where the training or education occurred is really of no importance to most vehicle owners. It is, however, critically important to you and me—or at least it should be. Our success or failure is the direct result of our skill and ability, or the skill and ability of our technicians. Yet we seem unable or unwilling as an industry to charge more for a technician who is schooled and certified in the necessary disciplines. How or why should our customers value that skill if we do not?

Ability

Ability goes hand-in-hand with skill and is equally ignored. The bottom line is clear—skill and ability are certainly a part of the package the customer is interested in, but they are assumed and therefore expected. If you insist on continuing education for your technicians—if you send them to school and see that they are properly prepared to meet the challenges your customers' vehicles can produce—aren't they worth a few dollars more per hour than the guy down the street who has repeated his one year of experience for the past twenty years?

Ethics, Honesty, and Morality

Ethics, honesty, and morality are givens as well. Every motorist has made a leap of faith that assumes the person to whom he has brought his car or truck is ethical, honest, and moral, despite the fact that they think the rest of us are crooks, thieves, and idiots! If you don't believe me, ask yourself this question: Would any reasonable, rational adult take their vehicle to someone they knew to be a thief? The answer is, of course not! What is it worth to the customer to know they are in the hands of an honest and ethical professional when they are buying a service about which they have no knowledge and no way to judge the full value—especially when newspapers, radio, and television are filled with stories of automotive service rip-offs?

Security and Confidence

Peter Revson, heir to the Revlon cosmetics fortune, once said in an often-quoted speech that women don't buy cosmetics, they buy hope. They purchase the illusion of beauty for themselves as manifested by someone else. Vehicle owners are not very different. In general, they are not interested in time, tools, technology, or any other facets of our business. In most cases, they couldn't care less about any or all of the above! They are interested in two things and two things only—their sense of security and their confidence in the vehicle.

Vehicle owners do not want batteries; they want cars and/or trucks that start! They do not want or need air conditioning maintenance, service, or repair. They want COMFORT! They want to get in their vehicle, regardless of brand, age, or previous maintenance, and get to where they are going.

Time, specialized tools and equipment, high-tech/high-touch, competence, skill, ability, and everything else we've discussed are just some of the forces we have to marshal to succeed in giving the vehicle owner what he wants. But none of them will take the place of the confidence and security the owner *needs*. To prove this, ask yourself one question: When do customers finally get rid of the cars and trucks that have served them so well—the ones you worked so diligently to restore? In most cases, it is after their vehicles have let them down once too often. When a motorist feels betrayed by his vehicle or the shop responsible for keeping it running, the vehicle is likely to end up in someone else's driveway and the customer is likely to end up in someone else's waiting room!

Success in business begins with knowing your product and knowing your customer. Your product is the security and trust we have been talking about. It comes from a number of things; however, in my opinion, it is primarily the result of only two of them—a constant pursuit of technical excellence coupled with achieving an intimate knowledge your customer's wants, needs, and expectations. That is what I mean when I say, "Know your customer."

You cannot deliver the kind of security and confidence your customer wants, needs, and expects from you without first recognizing the vehicle owner's need to believe in you and your ability to keep him mobile. Once you begin to see your business through your customer's eyes, sharing his feelings and emotions, success is only a few short steps away.

TRUST

The kind of security and confidence we have been talking about is deeply rooted in trust—the kind of trust that comes from delivering what you have promised. It comes from having the vehicle ready when promised, or (if it cannot be ready on time) from being sensitive enough to your customer's situation to call early enough that he can make other arrangements. It can also come from fixing the car and solving the motorist's problem on the first attempt each and every time. Unfortunately, performing in a service environment such as ours can complicate this relationship significantly.

You see, when a vehicle owner comes to you for automotive service for the first time, they give you their trust (something they value highly) *before* you deliver on your promise to perform. We forget that all too often. We actually owe the customer great performance the minute they step through the door—especially if they have never patronized our business before—because they have initiated the interaction and offered something of value first.

You may have invited them in with your marketing and your advertising, but they did not have to accept the invitation. It is therefore critically important to understand why we use the word trust so often. I think we have to understand how fragile it is—how easily damaged. I think we have to know what it means to the customer, especially when he offers it so freely, and I think we have to know what it means to us. In order to do that, we have to define it.

Trust rests on five pillars: predictability, accountability, reliability, response-ability (responsibility), and integrity, as seen in Figure 3-1. Without all five of these pillars to support the weight of the motoring public's trust, the security we have worked so hard to build will collapse around us. With these supports, we are almost ensured success.

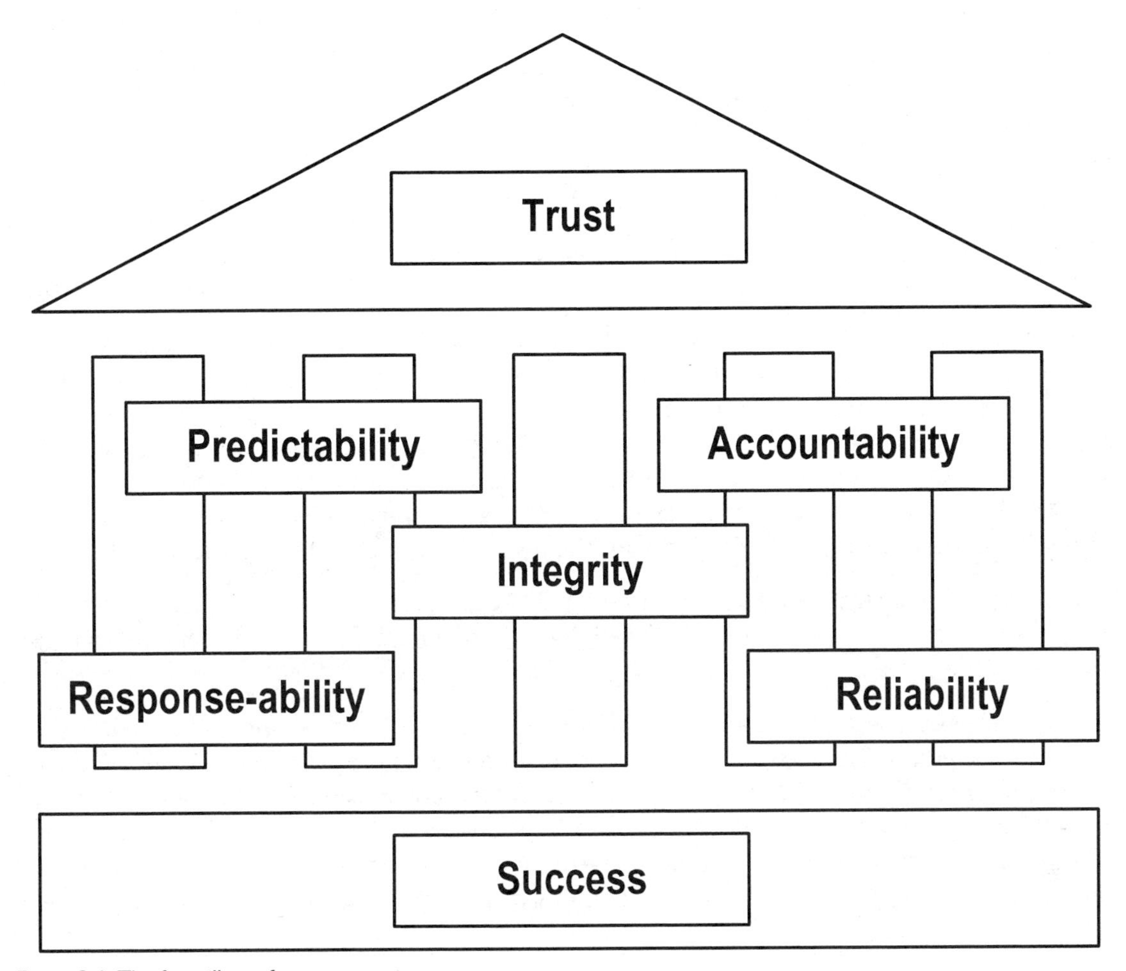

Figure 3-1 The five pillars of trust.

Predictability

Predictability can be defined as the sense of security your customers have in knowing how you will react to almost any situation. Fortunately (or unfortunately), it is developed through multiple interactions and takes time. It is, however, invaluable in the development of trust. The vehicle owner has to know that you will be there for him when he has a problem. He has to know that you will do what you said you would do. Predictability insures that if or when there is a problem, you will listen first and act later, inserting a space between stimulus and response—a space large enough for logic and reason, and sometimes compassion and understanding, to exist simultaneously. Predictability goes hand-in-hand with dependability.

Accountability

Accountability ensures that someone will be there to listen to the customer if or when there is a problem or a question that must be answered. Accountability denies the kinds of statements we hear all too often in any service interaction—statements like "There's no one here to answer that question for you now," or "Gee, I don't know," or "No, I'm sorry, I can't help you."

The automotive service aftermarket exists because of the vehicle owner's perception that there was no accountability wherever the previous choice for service was made.

Reliability

Reliability is different than predictability. After all, we all know people who are predictably unreliable! The motorist is not likely to tolerate unreliable performance for very long—especially if he feels someone more reliable is available down the block or around the corner to solve his problems.

There is an old joke about two hunters camping in the woods who were awakened early one morning by a bear rampaging through their camp. One of the hunters jumped out of bed and began running as fast as he could while the other stopped to tie his shoes. The first hunter called back to his friend in disbelief wondering what in the world he was doing at a time like this. He replied that he stopped to tie his shoes so he could run faster. The first hunter yelled back that he didn't think the second hunter could outrun the bear. And the second hunter yelled back as he gained on his friend that he did not have to outrun the bear. All he had to do was outrun HIM!

You don't have to be faster or better or more capable than everybody, you just have to shine in comparison to everyone else in your little corner of the marketplace—at least everyone trying to compete with you!

The vehicle owner wants to know that when his vehicle is in for service, maintenance, or repair, competent, well-trained, and professional technicians will perform the necessary work on his vehicle. He also wants to drive away knowing that the benefit of that service will be a car that starts at midnight or a truck that does not leave him stranded on the side of the highway. In order for that to happen, the service facility that he has chosen needs to respect the fact that he did have a choice in automotive service and respond by delivering first quality service each and every time it is called upon to do so.

Responsibility

Responsibility is really two words—response and ability—joined together to create one singular concept. The first word (response) insures that someone in your organization is willing to meet the challenges that delivering high quality service entails. The second word (ability), speaks for itself. Response-ability means listening to the owner when there is a problem, then responding appropriately. That is why I have hyphenated responsibility, because at that point responsibility becomes response-ability—the ability to respond to the wants, needs, and expectations of the customer positively and quickly. It recognizes the need to become response-able—able to respond and to meet or exceed the customer's expectations.

Integrity

Integrity is not generally included in discussions of trust, although in my opinion trust cannot exist without personal or professional integrity. The same root for the word integrity is found in the word integrated, and both suggest *oneness* or wholeness—a commitment to one's core values or one's sense of self-worth and self-esteem.

Trust is the dividend that you receive when you exercise integrity, and integrity grows when you recognize that you are worthy of trust.

Without predictability, accountability, reliability, responsibility, and integrity, trust cannot exist. Without trust, neither leaders nor leadership can exist because trust is the foundation of leadership. Without leadership, your business will flounder and ultimately fail, and without leaders this industry will surely fail.

CHAPTER SUMMARY

You and I sell time, specialized tools and equipment, high-touch/high-tech services, a physical plant, competence, skill, ability, ethics, honesty, morality, security, and confidence to the customer. Security is provided to the customer through the five pillars of trust—predictability, accountability, reliability, responsibility, and integrity.

How will you and your company continue to enhance your relationship with the customer? Write your thoughts in the chart on the next page.

Thoughts

Actions

Results

CHAPTER 4

Customer Relations

INTRODUCTION

This chapter introduces you to:

- What the customer expects from us
- What the customer has no right to expect
- How to manage your expectations

WHAT IS THE PURPOSE OF BUSINESS?

Peter Drucker once said that "the purpose of business is to find a customer!" Theodore Levitt of Harvard University amended that famous statement to read "The purpose of business is to find *and keep* a customer!" That directive probably ought to be amended still further to read "The purpose of business is to find and keep *the right kind* of customer!"

All that remains is to figure out HOW to accomplish that, and for the next few pages, that's exactly what I'd like to do. Let's talk about customer expectations, customer satisfaction, customer loyalty, customer retention, and finally, just what a relationship with this *right kind of customer* is supposed to look and feel like. Customer relations encompasses just about any of the interactions that take place between you, your company, and the customer, and that's virtually everything related to your business.

Customer Expectations

Customer expectations are at the very heart of our relationship with the vehicle owner. In fact, there can be no discussion of a relationship with the owner until we first understand what that owner wants, needs, or expects from us.

Quality Parts & Service

The vehicle owner expects to receive the best quality parts and service available when she enters the automotive service environment—in many cases, more than she is willing or able to pay for. Reconciling the difference between what she expects and what she is willing or able to afford is more an issue of *managing* those expectations, however, than it is an issue of understanding what they are.

Competent & Qualified Professionals

She expects to have the work performed by competent and qualified professionals schooled in the latest technology and working with cutting edge equipment in a safe, clean, and professional environment. Nothing less makes any sense to the customer—nothing less should make any sense to you either.

Time

She expects someone to take the time and make an effort to explain—in terms she can understand—exactly what needs to be done, why it needs to be done, and how doing it will benefit her. She expects to be told how much it will cost to do whatever needs to be done and how long it should take to complete that work. But more than that, most customers expect to be informed if the repair is going to cost more than they were originally told or if it will not be ready when it was promised.

Written Estimate

If the customer does not expect a written estimate explaining exactly what she is buying, what is to be done, and the cost involved, she should! No one should have to enter into any service agreement without knowing what to expect from the other party, and that includes you and I. Furthermore, the customer should expect to be notified of any changes to that original agreement and should trust that nothing other than what was agreed upon will be done without additional conversation and authorization from her.

Guaranteed Work

Whether she asks you about it or not, the customer expects you to guarantee the work you have done and the parts you have installed—probably for longer than is realistic or reasonable. But that doesn't change her expectations, so you should manage that expectation by handling it up front and in writing so it is clear to everyone.

Information

If it becomes apparent that additional inspection and testing is necessary, which is often the case, the customer expects you to help them understand how and why that work is important. More than that, they want to know what you hope to accomplish as a result of what you learn, how much it will cost, and how it will benefit them. It is the customer's vehicle and the customer's money, so she is entitled to information that she can process, written clearly and legibly on an invoice that she can read, with both parts and service operations identified and itemized.

Knowledge and Trust

Most customers expect to have their questions answered—even the ones you and I might think are foolish or stupid. The customer expects to be able to trust you, your people, and the people you do business with. Trust is what the customer-service provider relationship is based on, and managing these expectations honestly, successfully, effectively, and efficiently is what success in this business is all about!

WHAT THE CUSTOMER HAS NO RIGHT TO EXPECT

A vehicle owner should expect trust, honesty, integrity, performance, quality, service and much more every time she enters the automotive service arena. In my opinion, however, there are some things the consumer has no right to expect. The customer has no right to expect more automotive service than she is willing or able to pay for. There is no such thing as a free lunch and if the person you are dealing with is unwilling to allow a fair return on your investment, you don't need to participate in a relationship that is unhealthy, unprofitable, or self-destructive.

The customer *should* expect to be treated with courtesy, consideration, and respect. Nothing less is acceptable in any professional relationship. By the same token, however, they must understand that you as the service provider are entitled to the same courtesy, consideration, and respect! She has purchased your time, your tools, your skill, your ability, and your expertise—not your dignity and self-respect. You should expect the same respect afforded to any other qualified professional.

Customers should not expect you to compromise your integrity. They have no right to ask you to say or do anything unethical, dishonorable, or immoral.

While the customer has the right and the responsibility to define her expectations with regard to service, I have the same rights and responsibilities to define what I expect from a good customer.

MANAGING EXPECTATIONS

The key to successfully managing these expectations on both sides of the service counter is a special kind of consistency born out of policies, processes, and procedures with meeting or exceeding the customer's wants, needs, and expectations at their center. Managing these expectations begins with your image and the customer's awareness of both you and your facility. It is the direct result of each impression the customer has experienced regarding your business—known or unknown, recognized or unrecognized, conscious or subliminal. All of your marketing efforts, direct or otherwise, either create or reinforce that image, and position your business and all of the products and services you offer the motoring public in the mind of the consumer.

Public Relations

It could be a direct mail marketing piece, a customer reminder post card, an advertisement in the local newspaper, or a thirty-second TV spot on the local cable channel. It could be word of mouth at a local church social, an automotive literacy class for the local Girl Scout troop, or a billboard in left field at the high school baseball field.

Building credibility is the sum total of your public relations efforts. In the same respect, building visibility is the sum total of your advertising efforts. It is everything you do to bring a first-time customer to your door or help convince an existing customer to return. Each piece of this puzzle reinforces the customer's awareness and reinforces your image in that customer's mind. It is that position—that direct relationship between automotive service and Schneider's Automotive Repair (or whatever you have named your business)—that we all fight so hard to reach. It is that position that is so easily destroyed if experience does not match expectations.

Honesty

One of the critical ingredients for success in our business—or any business, for that matter—is honesty, especially when it comes to being honest with ourselves. Never promise anything you can't deliver—not in your advertising and marketing, and not in the service bay. In fact, the best advice anyone will ever give you when it comes to service performance is *under-promise, and over-deliver*!

Don't promise a clean, warm, and inviting customer lounge if your customer lounge isn't clean, warm, and inviting. The worst thing you can do is invite someone for an experience that doesn't even come close to meeting their expectations! Don't promise prompt and courteous service if no one in your organization knows how to deliver it, or if you can't or won't deliver it consistently. Never promise that the vehicle will be ready at a certain time and then deliver it an hour late! Regardless of the reason, you have failed the customer at the most basic level of performance. And finally, don't exceed your estimate if there is any way you can avoid it—not without calling for additional authorization first!

Communication

What you can do is develop the policies and procedures necessary to insure that everyone is communicating and communicating clearly. You can manage your customer's expectations by communicating early in the process about what is reasonable with regard to automotive service. You can ask her when she expects the vehicle to be completed, knowing full well that every time you ask, she is likely to answer with an expectation that has no basis in reality. You can then help her understand what really is reasonable. It won't be long before *same day* automotive service is a thing of the past. You might as well start getting her prepared for that inevitability by letting her know that appointments are made for the whole day and that the automotive service process begins with inspection and testing, moves toward a diagnosis, then to an estimate, and finally ends with the necessary service or repair.

Managed Technician Performance

In order to manage customer expectations properly, you must manage technician performance. In order to do that you must communicate just as clearly and well with your technical staff as you do with your customers. You will never be able to complete the vehicle on time if the technician working on the vehicle has no idea when you promised it would be finished. You will never be able to monitor productivity if the technician has no idea how much time was allotted for each individual service operation or for the job as a whole, or how much time was actually spent working on the vehicle. The office staff will not be able to communicate clearly, honestly, or effectively with the customer if the service staff cannot (or will not) communicate with the office staff.

In the volume of work to follow, I will explain the system we use here at our shop. It is not the only system available designed to enhance service operations, but it is the one we've found that works best for us. It was specifically designed to manage the flow of information through the shop, to enhance communication (especially non-verbal communication), which increases efficiency and productivity while reducing the likelihood of chaos. The result, for us at least, has been a much greater ability to meet or exceed our customer's wants, needs, and expectations without the stress and frustration we experienced regularly before we began using it.

Identifying Customer Expectations

Realistically, managing customer expectations begins with identifying exactly what those expectations are. It begins with information gathered when the vehicle is brought in for service or (ideally) when the initial phone call for information, service, or an appointment is made. Information-gathering like this is invaluable. It helps you *qualify* first-time customers as well as better serve existing customers. It helps you become proactive with regard to this relationship because it gives you the opportunity to order parts and/or marshal technical service resources *before* the vehicle arrives in the shop. It gives you the opportunity to look at the technical literature and scan for technical service bulletins or recalls related to the symptoms identified by the owner. On factory-required and scheduled maintenance, it gives you the opportunity to pre-order the parts you might need to *fix the car right the first time* and still have it *ready when promise*d. It also helps you familiarize yourself with what the customer wants, needs, and expects from you and your staff.

Customer Requests

We break this process down into three components for each service request or concern on the repair order, and this is how it helps manage customer expectations. The customer expects to get his or her vehicle serviced or repaired properly the first time it is in for service. You and I both know this is next to impossible if the problem is not communicated properly—translated accurately from customer to technician and then back again. We do that by dividing service requests into two separate categories. The first is the *customer request.* This is the service or a repair the customer asked for when the vehicle was brought in for service, as seen in Figure 4-1.

The customer saw a leak in the upper radiator hose and *requested* that it be replaced. In our shop, that's exactly how it would be written up on the repair order, as depicted in Figure 4-2.

The customer request is pretty straightforward and really doesn't require a lot of explanation. It's important to remember, however, that regardless of what the customer asks for, you and I are the professionals in this relationship, we are compelled to do what is right—to act in the client's best interest even if she refuses to. In other words, don't replace a part that does not need to be replaced until you have communicated the possible consequences of her request.

Customer Concerns

The other category of the service request is *customer concerns.* A customer concern is a rattle or a thump, a clank or a bang, or a green, brown, pink, or clear liquid under the vehicle. It is air conditioning that won't blow cold or a heater that won't blow hot. It is fuel economy that has mysteriously slipped, or performance that has disappeared. It is an intermittent anything. It isn't as clean and neat as a customer request, but for most shops it describes the bulk of the service work that we are asked to do. At our shop, it would look something like what is illustrated in Figure 4-3 and Figure 4-4.

That may seem like an unnecessarily lengthy explanation, but automating the process with canned

607 LOS ANGELES AVE • SIMI VALLEY, CA 93065
(805) 581-2340 • FAX (805) 581-5679
B.A.R. #AF-1154154

ESTIMATE W000021788 DATE 27 Jun 2002 TIME 11:51:13

Schneider's Auto Repair
607 Los Angeles Avenue
Simi Valley, Ca
93065
Home(805) Bus(805) 581 2340

Lic: WO #: W000021788
VIN:
Make: Chevrolet Year: 1985
Model: S10 Pick-U Unit:
Colour:

LABOR DESCRIPTION	PRICE	TAX	QTY	PARTS DESCRIPTION	PRICE	TAX

Estimated at $53.53 (incl.taxes) on 06/27/02 at 11:51.
Odometer reading 3475.......................................

IF POSSIBLE, WHEN WOULD YOU LIKE THE VEHICLE
COMPLETED ____________
IF FOR ANY REASON THE VEHICLE CANNOT BE
COMPLETED AT THAT TIME, WHEN WOULD YOU LIKE TO
BE NOTIFIED__________AM/PM ()......................$0.00

CUSTOMER REQUEST: UPPER RADIATOR HOSE IS
LEAKING @ THE WATER OUTLET HOUSING CONNECTION
REMOVE & REPLACE HOSE AS REQUIRED ()..............$50.98

Figure 4-1 Customer's estimate copy with customer request.

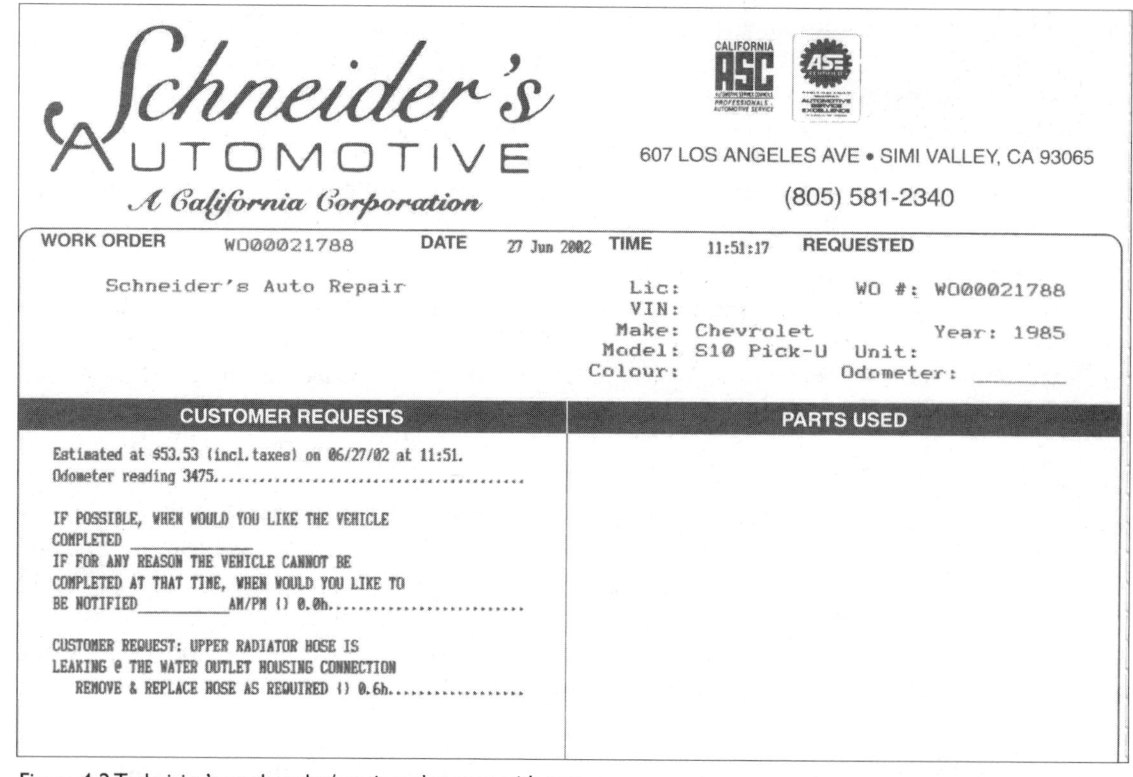

Schneider's AUTOMOTIVE
A California Corporation

CALIFORNIA ASC

607 LOS ANGELES AVE • SIMI VALLEY, CA 93065
(805) 581-2340

WORK ORDER W000021788 DATE 27 Jun 2002 TIME 11:51:17 REQUESTED

Schneider's Auto Repair

Lic: WO #: W000021788
VIN:
Make: Chevrolet Year: 1985
Model: S10 Pick-U Unit:
Colour: Odometer: ________

CUSTOMER REQUESTS	PARTS USED
Estimated at $53.53 (incl.taxes) on 06/27/02 at 11:51. Odometer reading 3475....................................... IF POSSIBLE, WHEN WOULD YOU LIKE THE VEHICLE COMPLETED ____________ IF FOR ANY REASON THE VEHICLE CANNOT BE COMPLETED AT THAT TIME, WHEN WOULD YOU LIKE TO BE NOTIFIED__________AM/PM () 0.0h......................... CUSTOMER REQUEST: UPPER RADIATOR HOSE IS LEAKING @ THE WATER OUTLET HOUSING CONNECTION REMOVE & REPLACE HOSE AS REQUIRED () 0.6h.................	

Figure 4-2 Technician's work order/repair order copy with customer request.

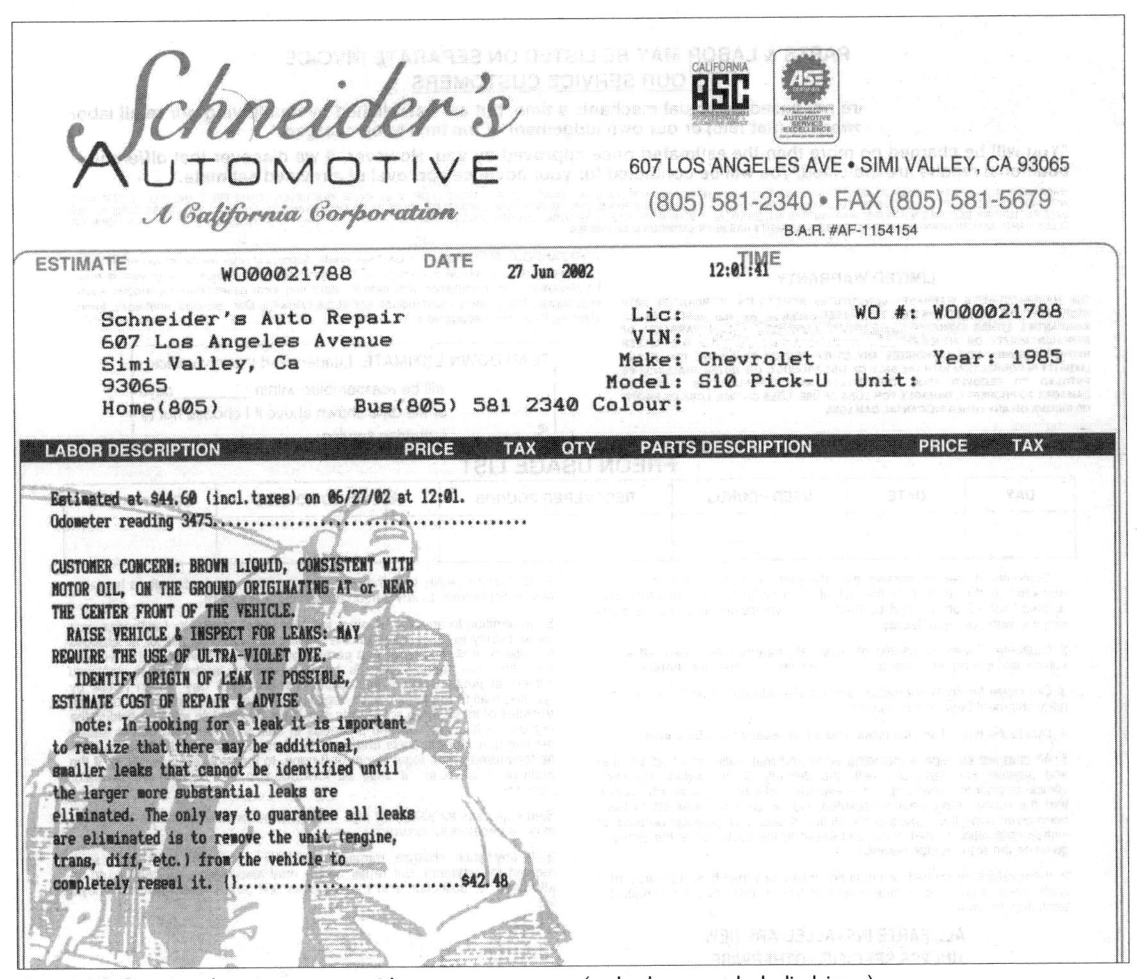

Schneider's
AUTOMOTIVE
A California Corporation
607 LOS ANGELES AVE • SIMI VALLEY, CA 93065
(805) 581-2340 • FAX (805) 581-5679
B.A.R. #AF-1154154

ESTIMATE W000021788 DATE 27 Jun 2002 TIME 12:01:41

Schneider's Auto Repair
607 Los Angeles Avenue
Simi Valley, Ca
93065
Home(805) Bus(805) 581 2340

Lic:
VIN:
Make: Chevrolet
Model: S10 Pick-U
Colour:

WO #: W000021788
Year: 1985
Unit:

LABOR DESCRIPTION	PRICE	TAX	QTY	PARTS DESCRIPTION	PRICE	TAX

Estimated at $44.60 (incl.taxes) on 06/27/02 at 12:01.
Odometer reading 3475..

CUSTOMER CONCERN: BROWN LIQUID, CONSISTENT WITH MOTOR OIL, ON THE GROUND ORIGINATING AT or NEAR THE CENTER FRONT OF THE VEHICLE.
RAISE VEHICLE & INSPECT FOR LEAKS: MAY REQUIRE THE USE OF ULTRA-VIOLET DYE.
IDENTIFY ORIGIN OF LEAK IF POSSIBLE,
ESTIMATE COST OF REPAIR & ADVISE
note: In looking for a leak it is important to realize that there may be additional, smaller leaks that cannot be identified until the larger more substantial leaks are eliminated. The only way to guarantee all leaks are eliminated is to remove the unit (engine, trans, diff, etc.) from the vehicle to completely reseal it. ()..............................$42.48

Figure 4-3 Customer's estimate copy with customer concern (and subsequent leak disclaimer).

labor operation codes and a computerized automotive shop management system can actually help speed things up. Regardless, taking a few seconds to explain what reasonable expectations are when the vehicle is dropped off can save you hours at the end of the process when you are dealing with a frustrated and disappointed customer who thinks you have failed because there are still a few brown spots under the vehicle after it's been parked for awhile.

By establishing up front that there could be other leaks, you have prepared the customer for that possibility. If it doesn't happen, you're a hero—but if it does, you aren't the scapegoat. By preparing the customer for the possibility of a leak, you have eliminated the expectation that there won't be. You have also given that customer an alternative, albeit an expensive one, to searching out and eliminating these leaks one at a time.

Cause

In either case, you have clearly identified exactly what it is you expect the technician to look for. The more clearly defined that task is, the less time it will take the technician to figure out exactly what you want her to do. The less time you waste thinking about what to do, the more time you have available for actually doing it. At that point, it becomes the technician's respon-

607 LOS ANGELES AVE • SIMI VALLEY, CA 93065

(805) 581-2340

WORK ORDER W000021788 DATE 27 Jun 2002 TIME 12:01:45 REQUESTED

Schneider's Auto Repair

Lic: WO #: W000021788
VIN:
Make: Chevrolet Year: 1985
Model: S10 Pick-U Unit:
Colour: Odometer:

CUSTOMER REQUESTS	PARTS USED
Estimated at $44.60 (incl.taxes) on 06/27/02 at 12:01. Odometer reading 3475...................................... CUSTOMER CONCERN: BROWN LIQUID, CONSISTENT WITH MOTOR OIL, ON THE GROUND ORIGINATING AT or NEAR THE CENTER FRONT OF THE VEHICLE. RAISE VEHICLE & INSPECT FOR LEAKS: MAY REQUIRE THE USE OF ULTRA-VIOLET DYE. IDENTIFY ORIGIN OF LEAK IF POSSIBLE, ESTIMATE COST OF REPAIR & ADVISE note: In looking for a leak it is important to realize that there may be additional, smaller leaks that cannot be identified until the larger more substantial leaks are eliminated. The only way to guarantee all leaks are eliminated is to remove the unit (engine, trans, diff, etc.) from the vehicle to completely reseal it. () 0.5h................................	

Figure 4-4 Technician's work order/repair order copy with customer concern (and subsequent leak disclaimer).

sibility to satisfy the request or address the concern. In other words, the technician's responsibility is clear—to identify what caused the concern or prompted the request. We call that *cause.*

Cure

Once the cause is identified, the solution to the problem can be identified, accurately estimated, communicated to the customer, and the additional service work can be authorized and cycled back into the production loop. Based upon the other work in the shop, how the work is loaded, and the availability of parts, a revised completion time can be calculated and the customer can be advised of the new time and the additional cost.

The results? No five o'clock surprise! No one glaring at you from across the service counter! No stress!

CHAPTER SUMMARY

To manage customer expectations successfully, we have to be aware of the different processes at work in the shop at all times. When the vehicle first comes in for service, we are looking at *exploration*—why is the vehicle here and what are we looking for? After that comes *explanation*—communicating to the consumer what we have found and the implications of those findings, both in time and in dollars. And finally, *authorization*—getting the customer's agreement to completing the process by finishing the service or completing the repair process.

For all this to happen successfully, mutual respect and excellent communication are required. Everyone must know what their role is and why that role is important to the overall success of the business. Process will define who does the inspection, the information required for a comprehensive, complete, and accurate estimate, and who makes the call to authorize additional service work. Thoroughness is required by everyone involved because no one likes to call the same customer more than once for an increase in the original estimate, any more than the customer likes to receive more than one phone call.

Managing customer expectations requires information, not manipulation. It demands that you and I understand that our role is to *facilitate* the buying process, not to sell by manipulating the customer or the situation. To most (if not all) of us, selling implies getting someone to spend money they don't necessarily have on something they may not want or need. Selling—aggressive selling—can often result in what is known as *buyer's remorse*... a phenomenon that occurs when the customer realizes that the decision to purchase may not have been entirely her own.

How will you and your company successfully manage customer expectations? Write your thoughts in the chart on the next page.

Thoughts

Actions

Results

Integrity-Based Sales

INTRODUCTION

This chapter introduces you to:

- What is best for the customer
- Sales
- Balance

WHAT IS BEST FOR THE CUSTOMER?

The key question in business is and will always be: What is best for the customer? The answer is Integrity-Based Sales (IBS), a system that has at its core these immutable values: integrity, character, service, and honesty. Value-based actions are rarely called into question. If and when they are, they are almost always self-explanatory. Take care of the customer's wants, needs, and expectations and you take care of yourself, your company, and your future. Like all systems, IBS is a process and as such can be institutionalized into your everyday operations.

IBS is discussed in great detail elsewhere in the Automotive Service Management series. However, I will quickly discuss some of the principles here as they impact customer relations and dramatically managing customer expectations. You serve the customer best by understanding what they want, need, and expect from you. In order to do that you must gather information—information about the client, information about the vehicle, information about the way it is used, how and where it is driven, who is driving, and what they want out of that experience.

You manage customer expectations with respect—respect for the vehicle, respect for the customer, and respect for the wants, needs, and expectations that flow from the kind of information gathering we have already discussed. To succeed, attention must be focused on the buyer.

SALES

Any time we talk about sales or the process of selling, where is the attention focused? Who are we really talking about? Who are we really talking to? Are we talking to, or focusing our attention on, the person doing the selling? Even the word, *selling*—a verb that describes an action—suggests that *someone* is doing *something* to *someone else*.

Think about the difference between taking an order, having a client or a customer tell you what they want you to do, and actively, aggressively selling something. When you sell aggressively, you use information to move the customer to where you want them to be—emotionally or psychologically.

When you focus on the wants, needs, and expectations of your customer, using your knowledge and experience to help that customer make the best decision possible for them, you may be active... but only in the pursuit of satisfying a customer and not a commission check!

BALANCE

The key is *balance*. The key is alignment with the customer at all times, not so that you can move, manipulate, or control, but so you can understand, evaluate, and suggest. It means not only walking away from a customer you know would be better served elsewhere—it also means suggesting the appropriate alternative, even if it puts that customer in someone else's service bay!

Alignment with the buyer also means understanding the various phases we all go through as we acquire information, evaluate what we've learned, and ultimately make choices.

If we believe that we can deliver the best quality automotive products and services available anywhere, we need to help the customer understand why we represent the best possible choice for those products and services.

CHAPTER SUMMARY

Integrity-based sales is a system that has integrity, character, service, and honesty as its foundation. By focusing on the wants, needs, and expectations of your customer, you will consistently be aligned with him.

How will your company help the customer understand the logic behind your business? Write your thoughts in the chart on the next page.

Thoughts

Actions

Results

CHAPTER 6

The Value Equation

INTRODUCTION

This chapter introduces you to:

- How the value equation works

HOW DOES THE VALUE EQUATION WORK?

Recognizing the importance of the customer's perception of value cannot be overstated. The difference between the perception of high value and the perception of no value is the difference between success and failure! Yet most of us haven't a clue when it comes to calculating or understanding how value is created. For years it was lost to me until the relationship between performance, cost, and value were finally explained to me, as seen here in Figure 6-1.

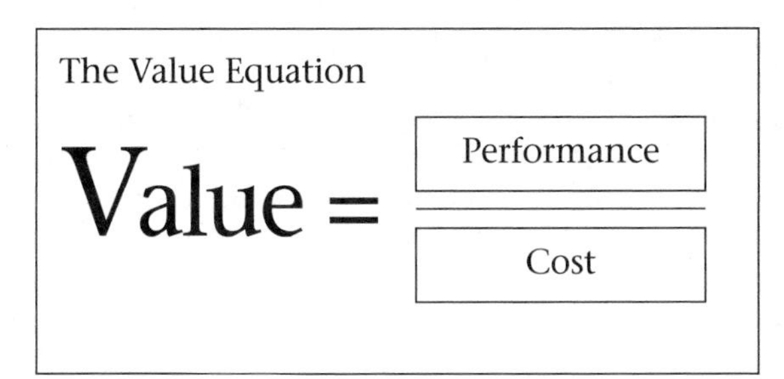

Figure 6-1 The value equation.

Value

val·ue (val'yoo) *Abbr.* val 1. An amount, as of goods, services, or money, considered to be a fair and suitable equivalent for something else; a fair price or return. 2. Monetary or material worth: *the fluctuating value of gold and silver.* 3. Worth in usefulness or importance to the possessor; utility or merit: *the value of an education.*

Value, in this case, refers to a somewhat different definition than the one used to denote a principle or a highly desirable trait. This definition of value is used to describe something of worth to the consumer—something the consumer cherishes, such as time, money, or prestige. It refers to something the consumer holds in high regard... something of significance... something desirable.

Performance

per·form·ance (pər-fôr-mĕns) *noun* 1. The act of performing or the state of being performed. 2. The act or style of performing a work or role before an audience. 3. The way in which someone or something functions. *synonyms*: working, functioning, execution, implementation... service

In other words, performance can be considered synonymous with service and service can be defined as an action giving aid or assistance to another or work done by one person for the benefit of another. Whatever it is, it is a verb—something referring to action. In other words, in order for it to be service, something has to happen... someone has to *do* something.

Cost

cost (kôst) *noun* 1. An amount paid or required in payment for a purchase; a price. 2. The expenditure of something, such as time or labor, necessary for the attainment of a goal. See synonyms for price.

Cost refers to the cost to the consumer in dollars or anything else the customer values—particularly time.

The presence of an equal sign (=) in the value equation indicates a mathematical equation—an equation that must remain perfectly balanced at all times. Change something on one side of the equation and it must be reflected or represented on the other. If you increase the cost of a service without increasing performance, the value must go down. Increasing performance while allowing the cost to remain the same drives the value up. Increase the cost and performance equally, and value will remain constant.

Understanding The Value Equation

How does understanding the value equation build customer loyalty? It does so by reminding us to remain constantly aware of what the customer perceives as good or beneficial to *her*. Whenever we increase the cost of a service to reflect our own increased costs—such as inflation or an increase in the cost of living—we must remember that without an equal or greater increase in performance (higher quality service) the consumer will feel as if they have actually *lost* something... as if something has been taken away from her. Increase the *perception* of performance by a factor greater than the increase in cost, and you will give the perception of greater or higher value.

Stated another way, if you give the consumer more of what she wants and needs, you will be able to increase the cost of your services enough to insure higher gross margins and greater profits.

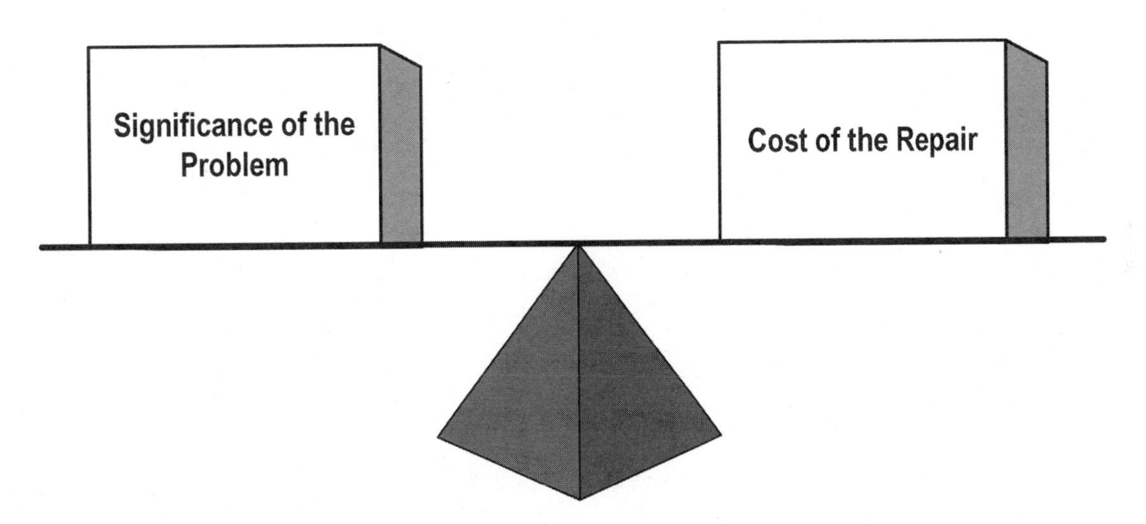

Figure 6-2 The decision to purchase.

Every purchasing decision is balanced on the head of a pin and the decision to purchase automotive service is no different. On one side you have the severity of the problem—on the other, you have the cost of the solution (see Figure 6-2) If the perception of the problem is not so great as to demand a solution, the cost of the solution will determine whether or not a purchase is made. If a motorist is driving down the highway with a rear wheel, anti-lock brake warning lamp on, but she still believes the vehicle is safe (whether it is or not), then there is no urgency to determine what is wrong and even less to repair the problem. If, however, the air conditioning compressor fails during a heat-wave in mid-August with 110° temperatures and 95 percent humidity, the *smart money* will be bet on a request for a quick fix regardless of the cost!

Remember, *if you can measure it, you can manage it! If you can't measure it, you can't manage it!* Using the value equation, we can do both. If we understand how the equation works, we can keep service quality high, insure higher margins, and still retain the perception of high value.

CHAPTER SUMMARY

Understanding the importance of a customer's perception of value means the difference between success and failure.

How will your company understand this important relationship? Write your thoughts in the chart on the next page.

Thoughts

Actions

Results

Customer Satisfaction

INTRODUCTION

This chapter introduces you to:

- Your relationship with the customer
- Customer satisfaction
- Our *high-demand* industry
- A new age
- Achieving customer satisfaction

OUR RELATIONSHIP WITH THE CONSUMER

We've discussed managing customer expectations, which is only one facet of our complicated relationship with the consumer. There are still the areas of customer loyalty, customer retention, and customer satisfaction that still need to be explored and understood. Finding success in only one dimension of our relationship with the vehicle owner is not enough to guarantee success. You and I must become proficient at managing all four areas. To successfully accomplish that, you must first understand how they work and how they work together. Then, and only then, can you begin to develop the metrics necessary to measure success.

We can start with the concept of *loyal and lifetime* customers and just what their value is to you, your business, and your future. I first came across the notion of a loyal and lifetime customer more than a decade ago in Carl Sewell's book *Customers For Life,* and it changed my professional life dramatically. At the time, Sewell owned the largest Cadillac dealership in the western United States and suggested that a loyal and lifetime customer was worth approximately $332,000 to his dealership: $300,000 in new car sales *and sales referrals* and $32,000 in service work *and service referrals.* Of course, his book was first published in 1990, so those values have certainly increased since then.

As the owner/manager of a small, independent repair shop in southern California, I can remember thinking to myself just how insane those numbers sounded! Even if you backed the $300,000 for new car sales and referrals out of the equation, the notion that a service bay customer could be worth over $30,000 seemed somewhat ludicrous. But then I started to look at the long-term relationships we have enjoyed with some of our clients here at the shop. We have families that have been trading with us for more than twenty years! We have worked on their vehicles from the time they were new, and as they were handed down from one son or daughter to the next. We have worked for them, for their friends and relatives, and even for their in-laws from time to time. That's when I made the connection. Twenty years of service work on one vehicle from between $500 to $1,500 per year, multiplied by the number of vehicles in the family, and multiplied by the number of referrals—and all of a sudden, that $32,000 figure didn't look quite so unbelievable!

Once you realize this, the question is: What can I do to maximize the relationship and ensure long-term success? What can I do to increase customer loyalty and keep customers for life? In order to understand customer loyalty you must first begin by understanding customer satisfaction, for there is no loyalty where satisfaction cannot be guaranteed.

CUSTOMER SATISFACTION

> *Marketers who build sturdy and enduring relationships with consumers act on the understanding that integral to successful product marketing is the establishment of an information rich customer environment characterized by exceptional ATTENTIVENESS, CONVENIENCE, ASSURANCE, and COMFORT—in other words... SERVICE.*
>
> *Regis McKenna, Real Time*

with regard to processor speed and cheap memory, micro-processors (computers) make formerly impossible business management, production, and manufacturing processes not only possible, but relatively easy. The combination of increasingly higher processor speeds and increased memory ensures that more work can be done in less time by fewer people (as seen in Figure 7-1)... Over the years, that has meant a certain amount of *displacement* (layoffs) or *readjustment* for many industries, leaving increased workloads for those workers remaining in the affected industries with few (if any) industries remaining immune to the onset of the Information Age.

Figure 7-1 Fewer people means more work.

As a result, *time* has consistently become one of our most critical considerations... and one of our most guarded and valuable possessions.

time (tīm) *noun Abbr*. t., T. **a.** A nonspatial continuum in which events occur in apparently irreversible succession from the past through the present to the future. **b.** An interval separating two points on this continuum; a duration: *a long time since the last war; passed the time reading*. **c.** A number, as of years, days, or minutes, representing such an interval: *ran the course in a time just under four minutes*. **d.** A similar number representing a specific point on this continuum, reckoned in hours and minutes: *checked her watch and recorded the time, 6:17 A.M.* **e.** A system by which such intervals are measured or such numbers are reckoned: *solar time*.

The economic dynamics of the Information Age have certainly resulted in more demands at work as work expands and the workforce contracts. Social and/or cultural dynamics at home have created equally stressful demands when it comes to personal time (family and recreation).

With time such an important consideration for everyone, people are increasingly less tolerant when their time is wasted or abused. In fact, I would suggest that wasted or abused time is what bad service and dissatisfied customers are all about.

ACHIEVING CUSTOMER SATISFACTION

Achieving customer satisfaction isn't really a mystery... It isn't accomplished with sleight of hand or distraction, misdirection, or smoke and mirrors. On the contrary, it is accomplished by understanding each customer's wants, needs, and expectations, and it begins with internalizing terms and concepts like quality, value, service, and satisfaction, and how they interact to insure that a *first-time* customer becomes a *loyal and lifetime* customer.

Quality

qual·i·ty (kwŏl-i-tē) *noun Abbr.* qlty. **1. a.** An inherent or distinguishing characteristic; a property. **b.** A personal trait, especially a character trait: *someone with few redeeming qualities.* **2.** Essential character; nature: *Mahogany has the quality of being durable.*

For our purposes, we will define quality as:

> The ability of a *product* or *service* to *meet* or *exceed* a customer's *wants, needs* or *expectations.*

By defining quality in these terms, we can measure not only whether or not our total quality efforts are successful, but just how successful they are. We can quantify and then control our ongoing quality control efforts, continuing to do what is working effectively and abandoning what is not.

As a purchaser of automotive parts and service, what are my criteria? How do I judge whether or not my money and time were well invested? First and foremost is this question: Did what I purchase do what it was purchased to do? If my goal was to purchase confidence and security, did I leave the shop feeling secure and confident that my automotive service needs were well met?

Let's judge this definition in clinical terms.

As a vehicle owner, I experience a disconcerting problem with my vehicle on the way to work. There is an acrid, almost burnt or burning odor, coming from under the hood, accompanied by some smoke. I pull into an automotive repair shop, Average Automotive, Inc., for help.

The people seem accommodating enough—they are polite, considerate, and concerned, which is not bad for the quality of service available in most retail environments across the country today. The facility is relatively clean and the office/waiting room seems professional enough, although a bit chaotic and disorganized. As I am describing my experience, the service writer/manager/advisor/owner is repeatedly distracted by the telephone and the questions of his other employees, and that is a cause for some concern. Regardless, I choose to leave the vehicle for inspection to find out how much it will cost, how long it will take, and when it will be ready.

Later on, I receive a call informing me that the valve cover gaskets are leaking oil onto the exhaust manifold and causing both the odor and the smoke. Oil is also leaking down the side of the engine onto the ground. A successful solution to the problem will require replacement of the gaskets. The cost is more than I expected, but the severity of the problem, in my mind at least, outweighs the cost of the solution and I agree to the repair.

What would make this a high-quality service? By definition, it would have to meet or exceed a certain set of wants, needs, and expectations, right?

As a vehicle owner, I want the vehicle to be ready when promised and fixed right the first time. That means picking up a vehicle that does not smoke, does not smell, and does not leak, at the promised time, and at the agreed upon price.

As a shop owner, it means a vehicle that does not smoke, does not smell, and does not leak, completed when promised, at a price that leaves profits adequate to meet my wants, needs, and expectations, as well as sustaining and ultimately *growing* the business.

If the service or repair costs less than expected, is completed before it was promised, and lasts longer than guaranteed (in other words, under-promised and over-delivered), we can all agree that the service was high quality. In other words, if the broken vehicle was restored to an acceptable baseline of performance, we have met the criteria for a high-quality repair or service. But is that enough? When we use words like *meets or exceeds* and *wants, needs, and expectations,* we are talking about the psychological components of automotive service as well as the mechanical requirements necessary to properly service, maintain, or repair the vehicle. Those are *warm and fuzzy* requirements that cannot be quantified so easily.

Therefore, to achieve any kind of substantive success we MUST understand just what the client wants, needs, and expects from us—wants, needs, and expectations that go beyond *delivered when promised* and *fixed right the first time* to security and confidence. That means spending time with the *broken person* to insure they understand what you are doing, why you are doing it, when the vehicle *should* be ready, why it isn't ready if it can't be delivered on time, what it's going to cost, and why the investment is justified.

Value

We will define value in terms of the customer's perception of benefit received for the work or service performed. Interestingly enough, it looks very much like the definition of quality, but there are two caveats. Take a look and think about what kind of a difference these make. Value is:

> the ability of a product or service to meet or exceed a customer's *wants*, *needs*, or *expectations*, at a certain *price*, and at a certain *time*.

Those two caveats—price and time—qualify and quantify the service *before* the quality component of value can be applied. Why? Because it is *possible* to have a high-quality product or service that has *NO value*! How? When there is no immediate need or the price is prohibitive.

Go back to the above example. What happens to our equation if the valve cover gasket is still leaking, but there is no smoke, no odor, and no evidence of oil on the ground? What is the value in fixing something that is not perceived to be broken? Value has to be looked at from the perspective of the consumer at all times and in relation to the value equation as explained earlier in this chapter. If the severity of the problem, as perceived by the consumer, does not outweigh the cost of the solution, there is no need for service, maintenance, or repair. At least, there is none in the mind of the person responsible for writing the check or signing the credit card invoice.

Service & Satisfaction

Words like service, value, quality, and satisfaction can distract or confuse us if we are not sure of exactly what they mean. Many times, the meaning can only be defined by the person or group using (or experiencing) the term and that leads us to a number of interesting questions. For example, who defines service in the automotive service environment?

I have asked that question of literally thousands of shop owners over the years and always receive the same answer: "The customer!" What would you do if I challenged that answer? What would you do if I suggested that it is NOT the customer who defines service in the automotive service environment?

For our purposes, the word *service* is a verb! That means in order for something to be defined as service something has to happen: someone has to *do* something. The only person who decides what will happen in an automotive service facility is, or at least should be, the owner or manager!

Does the customer determine when the facility will open and close? Not really. Does the customer determine what products and services will be offered, and when? No. Does the customer decide what those products and services will cost? Absolutely not.

The customer can, and often will, let you know if what you have decided to do is desirable or even acceptable, but choices like these are made by the shop owner—the dealer principal.

The shop owner defines and then determines what good service will be in his or her facility. The customer determines whether or not those choices work, whether or not they are satisfactory, and whether or not they create value for the customer!

It is important when we use terms like *satisfaction* and *value* that we realize they are similar. It is equally important, however, to realize they are not the same. Value occurs as a consequence of use. It can be *perceived* by the consumer of a product or a service, but not *appreciated* until that product or service is *experienced*. Until experienced, a product—and especially a service—is only a promise of something that may or may not occur in the future!

sat·is·fac·tion (sat'-is-fak'-shən) *noun* **1. a.** The fulfillment or gratification of a desire, a need, or an appetite. **b.** Pleasure or contentment derived from such gratification. **c.** A source or means of gratification.

Satisfaction, on the other hand, is a measure of something enjoyed or experienced in the past. You can't say that a dinner was *satisfactory* until you have gone to the restaurant, been seated, ordered, been served, and finally eaten the meal. At that point, you have the information and experience necessary to determine whether or not the experience left you satisfied and content.

By the same token, an automotive service experience cannot be considered satisfactory until it is experienced, and to most vehicle owners that means not only the initial performance of that service, but how long and how well it lasts.

CHAPTER SUMMARY

To measure success, you should be able to manage and understand how to achieve customer satisfaction in a high-demand industry and a new age. Customer satisfaction can be achieved by offering quality service and value, as seen in Figure 7-1. In order to do that you have to see, feel, and experience service from your customer's perspective. That is not always easy to do from your side of the service counter. It is important to develop your ability to empathize with your customers—to feel their frustrations and insecurities. Because all of us are or have been customers at one time or another, with customer service and satisfaction experiences that probably range from exceptionally positive to exceptionally negative, that shouldn't really be all that hard to do. Once you have done that, you may find yourself more willing to see and feel what the vehicle owner is likely to experience when in your care and custody.

Value and Satisfaction: The Differences

VALUE

- Is what the consumer desires from the product or service
- Is oriented toward the future independent of when it is used or consumed
- Is independent of any particular product or service offering or supplier
- Provides direction for the organization: What must be done to create value?

SATISFACTION

- Is the consumer's reaction to, or feeling about what was received
- Exhibits a historical orientation that is formed during or after use or consumption
- Evaluation is directed toward a particular product/service offering or supplier
- Provides a Report Card for the organization about how they are doing in their value creation efforts

Figure 7-1 Value/satisfaction: the differences (adapted from Woodruff and Gardial, *Know Your Customer*).

How will your company achieve success through customer satisfaction? Write your thoughts and feelings about quality and value, customer service, and customer satisfaction in the space provided on the next page.

Thoughts

Actions

Results

CHAPTER 8

Ten Measures Of Service Quality

INTRODUCTION

This chapter will introduce you to:

- How service quality can be measured in ten ways: tangibility, reliability, responsibility, competence, courtesy, credibility, security, availability, communication, and by understanding the customer
- Understanding and improving service quality

SERVICE QUALITY

service (sər'vəs) *noun* A customer-based or user-oriented function, such as technical support or network provision. **1.** Work done by one person for the benefit of another **2.** A helpful or beneficial act giving aid or assistance to another

"It Isn't Service Until Someone Does Something. Until Then, It's Just Rhetoric!" Janet Lowery

"If you can't measure it, you can't manage it!" and if you can't quantify it, you can't measure it! Let's look at ten specific things you can address to both understand and improve service quality.

Tangibility

Perhaps first and foremost are those things tangible—those things you can see, touch, and feel. Certainly, the appearance of the facility—the actual physical plant itself—as well as the condition and appearance of the equipment, personnel, and all the marketing and communication materials are tangible indications that someone cares enough to present a professional image. Is the building clean, freshly painted, neatly organized, and inviting? Does the shop appear well organized? Is the equipment clean and seemingly well maintained? Are the employees uniform in their appearance and do they appear in uniform?

While it is possible for a service facility to appear out of control, disorganized, and chaotic and still deliver professional quality automotive service, most people will not accept the risk necessary to find out. Consequently, how you look—how the facility appears to the outside world—is critical to managing the customer's perception of the quality of service you are able to deliver.

Reliability

Service quality must be reliable. The facility must be able to deliver the promised service dependably and accurately, which is just another way of saying *ready when promised* and *fixed right the first time*. You have got to be able to do what you say you will do! You have to deliver what you promise.

Responsibility

You must be responsive to the customer's wants, needs, and expectations... that means a sincere willingness to help customers and provide prompt service. It means, as we said earlier, being response-able—having everything it takes to get the job done. But, more than that, it means being almost anxious to respond as well.

Competence

Service quality will be judged by your competence, especially in the automotive service arena. While certainly *not* a commodity, competence is *assumed* by virtually everyone seeking automotive service. Yet the skills and knowledge required to deliver quality service—or high service quality—are rare at best. Anyone involved in the automotive service industry shares a responsibility to ensure that the knowledge and skill required to properly service and repair the vehicle is available when required. Anything less is dishonest at best, and fraudulent at worst!

Courtesy

Courtesy is certainly a measure of service quality. The respect, politeness, consideration, and friendliness experienced by the vehicle owner/customer/client at the hands of *everyone* on staff will color the service experience positively or negatively and effect the overall perception of quality in a powerful and emotional way.

Credibility

Your credibility—the ability of the consumer to trust you and believe you—will impact the delivery of service quality dramatically. Whether or not the customer believes you to be forthright and honest will have a profound effect on your ability to present what the vehicle requires in the way of maintenance and repairs. It will determine whether or not both the customer and the vehicle will get what they need and get the lowest cost of operation and the highest return on investment.

Security

Whether or not the customer walks away from this interaction with the security she both wants and needs is critical to the perception of high service quality. Without security and confidence there can be no perception of quality, because that is ultimately what the vehicle owner is looking for.

Availability

In our industry, availability (or access) is also an area of service quality that is measured by your customer base the first time (and every subsequent time) service is sought. Approachability and ease of contact are among the many criteria we all use when we judge the quality of a service experience.

Communication

The ability to communicate clearly, effectively, and in language the customer can understand is critical to a positive service experience. Communicating with the customer doesn't only mean explaining what you did, why you did it, and what it will cost. It also means listening, and listening is something few, if any of us, are really very good at! Listening—*active listening*—means listening with *empathy, curiosity*, and *patience*. Michael Eisner, CEO of Disney, indicated that perhaps one of the most self-destructive managerial traits is not a "lack of understanding some arithmetic table, not the lack of understanding what the information highway is, but the lack of understanding of why somebody is unhappy!"

Being able to place yourself squarely in the middle of your customer's experience is what *empathy* is all about. It allows you to feel what the customer is feeling. It is a critical component of service quality just as are curiosity and patience.

Curiosity, for our purposes, is the desire to understand more about a given circumstance or situation and the willingness to suspend judgment, prejudice, and preconceived notions of whatever you might encounter.

Patience is the ability to allow the customer to tell her story, no matter how painful or time consuming it may be. It is also the ability to ask the right questions, wait for the right answers and to hear what is not said just as clearly as what is said.

UNDERSTANDING THE CUSTOMER

The final component—understanding the customer—is all about empathy: your ability to both know and understand the client's situation, feelings, and motives. It's about recognizing the customer's wants, needs, and expectations, whether obvious or hidden, as well or better than they do. And once recognized, it's all about addressing those wants, needs, and expectations by satisfying them—or at least making every effort to try.

CHAPTER SUMMARY

There are ten ways to both understand and improve service quality: tangibility, reliability, responsibility, competence, courtesy, credibility, security, availability, communication, and understanding the customer.

How will your company integrate these ideas to improve service quality? Write your ideas in the chart on the next page.

Thoughts

Actions

Results

CHAPTER 9

Understanding The Customer

INTRODUCTION

This chapter will introduce you to:

- The beginnings of customer satisfaction
- The importance of obtaining loyal and lifetime customers

THE FINAL COMPONENT

Understanding the customer is the final component of measuring service quality, and it includes just about everything we have been discussing. It ultimately includes your willingness and ability to know and understand your customer's wants, needs, and expects.

MEASURING CUSTOMER SATISFACTION

Until recently, customer satisfaction generally went unmeasured. It was assumed that if you worked on someone's vehicle and they returned to your shop, everything was all right. The focus was on dealing with the broken vehicle; the broken *person* was ignored or generally went unnoticed. The service industry was considered to be transactional in nature, and consequently, the actual quality of the repair or service was far more important than the *perception* of quality or the relationships surrounding it. In other words, a trip to the shop was considered to be more of an isolated event by both the service provider and the vehicle owner than an intrinsic part of a somehow greater and ongoing relationship.

As the industry moved from singular transaction-based events to an intimate and relationship-based connection, measuring the client's perception of service quality took on new meaning and greater importance.

At first, service quality was measured internally in much the same way manufacturing companies measure quality—in defects per number of similar service events. Measuring service quality in this way is infinitely easier than measuring service quality in terms of value or satisfaction.

You can quantify, track, and measure whether or not a vehicle was delivered at the appointed time. If it was promised at three o'clock and delivered at five o'clock, the two hours it was late is pretty obvious. You can quantify, track, and measure whether or not the vehicle was fixed properly the first time just as easily—perhaps even more easily. If it came in, was inspected, tested and diagnosed correctly, and the repair did not warrant a return to the shop for the same or a similar problem, it would be easy to assume that the customer was satisfied and quality service was rendered. If, however, the vehicle is returned to the shop for the same problem a day, week, or month later, then determining how or why the problem was not solved correctly the first time is just a matter of relentlessly asking the right questions. We found, however, that measuring service quality in this way does not take into account the very human component of the interaction. How does all of this impact the customer?

Once you ask that question, you have started a journey where the ultimate goal is to achieve a better understanding of how the customer interprets and internalizes almost everything you do. Consequently, delivery of service must be measured both internally (how many times the phone rings before it is answered, how it is answered, and what is said, and whether or not the vehicle is ready when promised or fixed right the first time), and externally, (how and what the customer thinks and feels before, during, and after that service is delivered). That is a matter of understanding the difference between *actions* and *expectations*.

Actions are what *you* are expected to provide the customer as a service professional. *Expectations* are the many and different ways the customer believes he or she will benefit from those actions.

Just Our Way of Saying

Thank You

We want to be your first choice in automotive service. In order to do that, we need your help. Please take a moment to answer the short questionnaire below so we can "fine tune" the Schneider's Automotive experience.

As a way of expressing our appreciation, each month we award a 15,000 mile service (more than an $80 value) to the person whose returned questionnaire is chosen by random selection.

Thank you,

The Schneider Family

Check only one box per question.

How would you rate our service ❑ Excellent ❑ Good ❑ Fair ❑ Poor

	Yes	No
Was your vehicle ready when promised?	❑	❑
If not, were you notified of the delay?	❑	❑
Were you satisfied with the work that was done?	❑	❑
Was it within the estimate you were quoted?	❑	❑
Was the facility clean at the time of your visit?	❑	❑
Were all your questions answered?	❑	❑
Would you recommend us to a friend or family member?	❑	❑
Will you return?	❑	❑

Comments ______________________________

____

Optional:

Repair Order ______________ Date of Service ______________

Name ______________________________

Figure 9-1 Sample customer service index survey.

IS CUSTOMER SATISFACTION REALLY THE GOAL?

In any discussion of our relationship with the consumer of automotive products and services, we have to ask ourselves if the ultimate goal is *customer satisfaction*. To some degree, the answer has to be yes. But is that our only goal? If most customers are satisfied, they will return, right?

The answer is both yes and no, and that changes our answer to the: "Is customer satisfaction our ultimate goal?" question. The goal is to find and keep loyal and lifetime customers, and you can't create a loyal and lifetime customer without a high degree of customer satisfaction. Research has proven, however, that while most customers identify themselves as being *satisfied*, they aren't automatically *loyal*.

LOYAL AND LIFETIME CUSTOMERS

The goal is to develop repeat customers! The goal is loyal and lifetime customers! Establishing a process for delivering high levels of customer satisfaction is a proven method of taking first-time customers and turning them into loyal and repeat customers!

Why is this important? For a number of reasons... First of all, loyal and lifetime customers are already familiar with who you are and what you do. You don't have to spend quite as much time creating a relationship or establishing credibility. The cost of marketing to an established client base is lower. Your message and your investment isn't compromised or diluted by sending it to individuals who neither want nor need the products and services you provide. Lifetime and loyal customers spend increasingly more the longer they are loyal and lifetime customers. Furthermore, the cost of *maintenance* is less—loyal and lifetime customers are generally *low maintenance*, requiring far less time and effort to satisfy than first-time customers.

Fredrich Reichheld, in his book *The Loyalty Effect*, suggests that a five percent increase in customer loyalty can result in a 50 percent increase in gross profits!

CHAPTER SUMMARY

The automotive industry has recently recognized customer satisfaction as having greater importance to the outcome of a company than was previously thought. The perceptions of the client and the quality of service have taken on new meaning. If customer satisfaction becomes our goal, most customers will return—becoming loyal and lifetime customers.

How will your company achieve customer satisfaction? Write your thoughts in the chart on the next page.

Thoughts

Actions

Results

CHAPTER 10

Customer Loyalty

INTRODUCTION

This chapter introduces you to:

- Why some customers remain loyal to a company, while others do not

REASONS FOR CUSTOMER LOYALTY

Surveys show that five to nine percent of a business's customer base will disappear every year. Customers will move, marry, pass away, decide they don't like you for one reason or another, or experiment with another service provider just to insure you are giving them the best service possible. Evidence suggests that the average business in the United States today loses half its customer base every five years. That translates to about a 13 percent loss per year. Just to stay where you are with no gains and no losses means increasing sales to your remaining customer base and new customers by 13 percent!

To survive, the average business in the United States today must enjoy an 80 percent loyalty rate. To succeed, that loyalty rate must be increased to 90 percent or higher.

The 80/20 Rule

If 80 percent of your customers are loyal, that means 20 percent may not be loyal and might switch at the next purchasing opportunity. That might not seem like a problem, but if you do $500,000 a year in gross sales you may be in danger of losing $100,000 at any given moment if you are not focusing on customer loyalty and satisfaction issues.

The Eight C's Of Satisfaction

Success Depends Upon the Eight C's of Satisfaction:

Control: whose?

Convenience: whose?

Consistency: whose?

Comprehensive: nature of the product or service.

Comfort level: whose?

Current: current wants, needs or expectations.

Capable: are you qualified?

Changeable: will you adjust to meet changes in the market place?

CHAPTER SUMMARY

The average business in the United States looses half its customer base every five years. That same average business must retain an 80 percent loyalty rate to survive. To achieve customer loyalty and satisfaction you should understand the 80/20 rule—that is, 80 percent of your customers are loyal, while 20 percent may not be.

How will your company retain its customers? Write your thoughts in the chart on the following page.

Thoughts

Actions

Results

CHAPTER 11

Customer Satisfaction

INTRODUCTION

This chapter introduces you to:

- How to achieve great performance and customer satisfaction in ten steps:

  Making the commitment

  Knowing your customer

  Developing a service strategy

  Developing a good service structure

  Finding good people and treating them like people

  Motivating and training every employee

  Empowering your staff

  Obtaining the right tools and resources

  Providing exceptional service

  Renewing your commitment to service satisfaction

HOW DO YOU ACHIEVE CUSTOMER SATISFACTION?

There is a technology for everything, and achieving high levels of customer satisfaction is no exception. Therefore, achieving customer satisfaction does not have to remain a mystery. It is simply a matter of determining how you can best meet or exceed the vehicle owner's wants, needs and expectations. You do that with a commitment to *great performance* each and every time service is delivered.

What is Great Performance?

Great performance can apply to just about anything, in almost any arena. It is as much a philosophy of excellence as it is anything else. Reduced to its most basic elements, achieving great performance can be understood best when reduced to the following four steps.

Defining the Term

First, define exactly what we mean when we use the term *great performance*. Remember what we said about measuring and managing? If you can't measure it, you can't manage it—and if you can't define it, you will never be able to measure it at all. So what kind of performance are we looking for and how can we describe it so that the results of our efforts will meet the performance criteria we have established?

Identifying Obstacles

The second step in achieving great performance is determining what obstacles stand in the way. It is reasonable to assume that, once recognized and defined, success through great performance is just a matter of meeting the requirements of that definition. If we don't (or can't, or won't), there has to be a reason or reasons, and understanding what those reasons are is critical to moving forward.

Identifying Who Owns the Obstacles

The third step is trying to determine who *owns* the obstacle or obstacles. Simply stated, the only obstacles you have any chance at all of removing are your own! You can't remove someone else's. Let me give you an example. One of the biggest problems in the automotive service industry today is training—not so much the *availability* of the training as the unwillingness of the service industry to *participate* in training. There are far too many excuses on both the management and production sides of the industry.

Management doesn't want to make the investment for fear the technician will move to a shop down the street for a few dollars more an hour—the cost of the training is too great. I've heard the same words uttered from a hundred different shop owners in a hundred different cities across the country, "I can't allow my best technicians to go to training during the day because of lost revenue issues or loss of production and the evening classes are a waste of time!"

On the other side of the wall, the technician is too exhausted to go to training in the evening after working ten or twelve hours a day in a production shop environment. The cost is too great and the reward too small. There is no incentive to go... and the list goes on.

What would happen if I removed all the obstacles from the management side? What would happen if the shop owner paid for the training... all the training? What would happen if the shop owner not only allowed the technician to go to training, but encouraged it as well? What would happen if you, as the shop owner, removed all the obstacles that might be considered yours? What would happen if you did all that and your technician still won't go to educational and training events available in your area? Whose obstacles would stand in the way of achieving great performance then—yours or your technician's?

You can't force someone to attend class if they choose not to. Even if you can get them to go, you can't force them to get anything out of the experience. If the technician had a bad experience in school, it's not going to be easy to force him back to the classroom and more difficult to get him to learn.

Acting As A Coach, A Mentor Or A Guide To Remove The Obstacles

Ultimately, you can't remove someone else's obstacles. That leads us to the fourth and final component of achieving great performance: doing whatever you can do—whatever is necessary—to act as a *coach, mentor,* or *leader* to remove the obstacles. One of the best ways you can do that is to help the person you are leading, coaching, or mentoring understand not only what you are doing, but why you are doing it and how their contribution can help.

It's one thing to announce that everyone in the company is responsible for achieving customer satisfaction and quite another to identify the ten specific steps necessary to do that.

TEN STEPS FOR ACHIEVING CUSTOMER SATISFACTION

Step 1: Making the Commitment

Make the commitment to total quality service—*the ability of your company to consistently meet or exceed the customer's wants, needs, or expectations each and every time a product or service is delivered.*

This is something that must flow from the top of your organization down. The best strategies, goals, and objectives will fail without a good role model, and as the owner or the manager, that is a role you have accepted. You must consequently conform to a "do as I do, not as I say!" philosophy. You must model, coach, and reinforce the skills you expect your employees to perform, and that means you must know and understand what kind of results you are striving for and why the effort is critically important. To be successful, you must educate, train, and motivate your employees to change their behaviors and actions, and this process must be continually reinforced.

To do this you must create a service culture... a *corporate* or business culture in which world-class service delivery is an intrinsic part of your company's personality—a personality that has, at its core, the belief that high-quality service is critical to the success of the company.

This quality of service should become expected—a part of your company's history, myths, traditions, and legends. In fact, delivering high quality service flawlessly and effortlessly should almost be taken for granted.

The benefits of creating such an organization are manifold. First and foremost, quality of workmanship and materials plus quality of service delivery builds pride throughout the organization, especially when people realize they are accomplishing something on a regular basis that no one else thinks is possible! Aside from that, the improved quality and workmanship that develop as a result of that pride leads to higher productivity, higher profits, and still higher quality.

As you can see from Figure 11-1, increased service quality results in fewer comebacks, high morale and dedication, increased opportunities to charge premium dollars, and a dedicated customer base. This will result in much greater marketing effectiveness and higher sales revenues, as well as decreased costs and increased productivity. This, in turn, gives us what every business both wants and needs, and that is higher profits to satisfy the stakeholders—the owners and shareholders—and the financial wherewithal to drive the business forward into the future.

Figure 11-1 Interrelationship diagram.

Step 2: Knowing Your Customer

Who are these people that make our professional lives possible? Where do they go, what do they read, and what do they watch and listen to? What do they buy and why do they buy it? To be successful, you must be able to identify the wants, needs, and expectations of your key or target customers! In an age of relationship-based interactions, you can't afford the luxury of assuming you know the customer's wants, needs, and expectations better than he does. But, you will need to anticipate where to focus your efforts, recognizing that you can't meet or exceed your customer's wants, needs or expectations if you don't know what they are.

Gathering Data

You learn about your customers by gathering data and asking questions:

What is it like to do business with our company?

What bothers you most about the way we sell to you or perform service for you?

What works about the way we conduct ourselves?

How can we make it better?

What can we do to save you time or make these interactions easier?

What would be the ideal interaction in automotive service?

Who is the company you most like to do business with and why?

The Questionnaire, Letter, and Survey

A questionnaire might look like this:

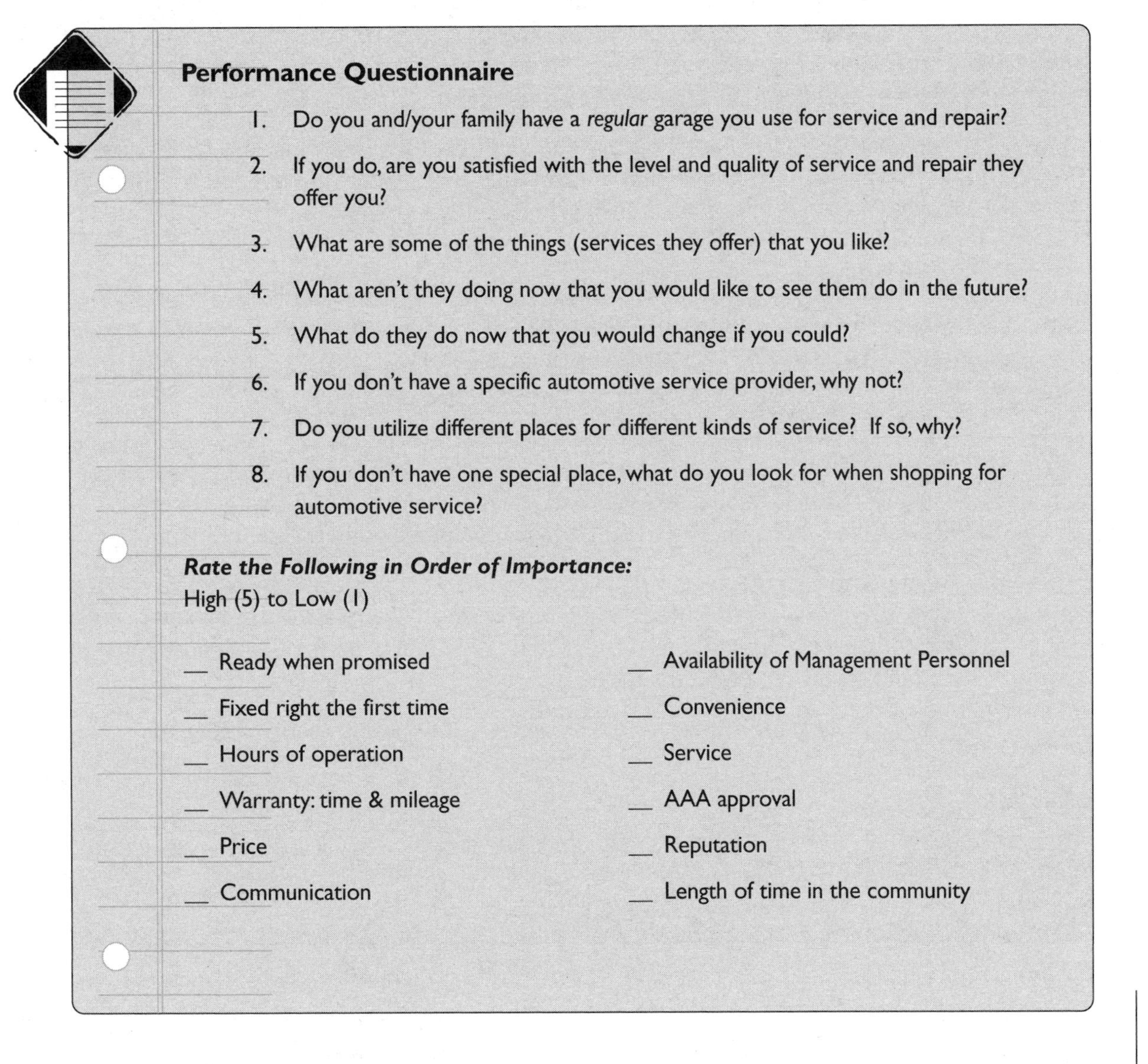

Performance Questionnaire

1. Do you and/your family have a *regular* garage you use for service and repair?
2. If you do, are you satisfied with the level and quality of service and repair they offer you?
3. What are some of the things (services they offer) that you like?
4. What aren't they doing now that you would like to see them do in the future?
5. What do they do now that you would change if you could?
6. If you don't have a specific automotive service provider, why not?
7. Do you utilize different places for different kinds of service? If so, why?
8. If you don't have one special place, what do you look for when shopping for automotive service?

Rate the Following in Order of Importance:
High (5) to Low (1)

__ Ready when promised

__ Fixed right the first time

__ Hours of operation

__ Warranty: time & mileage

__ Price

__ Communication

__ Availability of Management Personnel

__ Convenience

__ Service

__ AAA approval

__ Reputation

__ Length of time in the community

...and is accompanied by a letter like this...

Dear Simi Valley Vehicle Owner:

My name is Mitch Schneider and, together with my family, I own and operate Schneider's Automotive Repair. We are, and have been, consistently recognized as one of Ventura County's premier automotive service providers for over 20 years. To maintain that position of leadership in these changing times, we need your help.

We know that you have had the opportunity to form opinions and make judgments about the quality of automotive service available in our community. For us to do a better job serving of our customers, we need to know what those opinions are. We are asking you to share that information with us by filling out the enclosed questionnaire.

If you have a *regular* garage—a place you patronize for all your automotive service needs—we'd like to know what it is you like about what they are doing. What makes you feel comfortable there? What services do they offer that you feel are unusual, unique, compelling? Why do you keep going back?

We would also like to know how you think they could do a better job. What services you would like to see them offer that are not currently available today? And finally, if you are not a *regular* customer someplace, why aren't you? What would it take to make you a *regular* customer someplace... anyplace? What are you looking for?

We know that you, the customer, define what customer satisfaction is. By understanding what it takes to satisfy you, we can tailor the nature and quality of our service to meet the changing needs of both our community and the motoring public.

We also know your time is valuable. To compensate you for the investment in time, we will enter your returned questionnaire in a drawing—the winner and one guest will enjoy dinner on us (valued at $200) at one of Simi Valley's most popular and respected restaurants.

Thank you again for helping us to better understand the automotive service wants, needs, and expectations of our community.

Respectfully,

The Schneider Family

...and finally a survey that can be *quantified,* as seen in Figure 11-2.

AUTOMOTIVE REPAIR & MAINTENANCE

"GREAT AUTOMOTIVE PERFORMANCE" QUESTIONNAIRE

We want to be your best, most trusted and reliable auto service supplier.
To assure we accomplish this we need your thoughtful and critical input.
Please complete this questionnaire to help us help you.

1. What automotive repair or maintenance shops do you and your family regularly use now?

	Shops now used for different kinds of automotive services:		
Your Vehicles Yr. & Make	Diagnostic & Repair	Reg. Maintenance	Other Specific Svcs.
a. ______	______	______	______
b. ______	______	______	______
c. ______	______	______	______

2. How satisfied are you with the quality, reliability and need for the services above have provided?

a. 1 = not, 5 = very	1 2 3 4 5	1 2 3 4 5	1 2 3 4 5
b. 1 = not, 5 = very	1 2 3 4 5	1 2 3 4 5	1 2 3 4 5
c. 1 = not, 5 = very	1 2 3 4 5	1 2 3 4 5	1 2 3 4 5

3. What are three things you like most about the services and support you now receive from:

Shop Name:	______	______	______
What I like about them:	______	______	______
	______	______	______
	______	______	______

4. What are your three most important features or services of an auto service provider?
a. ______
b. ______
c. ______

5. What facts, features, services or things would cause you to select a new auto service provider?
a. ______
b. ______
c. ______

6. Please rate the importance to you of the following shop features or services (1 = low, 5 = high):

Person you deal with	1 2 3 4 5	Reputation or recommendation	1 2 3 4 5
Location of shop	1 2 3 4 5	Shop condition & appearance	1 2 3 4 5
AAA or other approval	1 2 3 4 5	Hours of shop operation	1 2 3 4 5
Credit card services	1 2 3 4 5	Charge account services	1 2 3 4 5
Tech communications	1 2 3 4 5	Management communications	1 2 3 4 5
Ready when promised	1 2 3 4 5	Fixed right the first time	1 2 3 4 5
Suggested future needs	1 2 3 4 5	New car warranty service	1 2 3 4 5
Return of replaced parts	1 2 3 4 5	Detailed repair order information	1 2 3 4 5

7. Would you have an interest in a service club membership with a small annual fee that provided:

- ☐ Four oil changes and lube services per year
- ☐ Regular inspection and recommendations
- ☐ Preferred scheduling for needed services
- ☐ Complete fluid check & refill on each visit
- ☐ Seasonal and safety checks as needed
- ☐ A car care partner who cares about you too!

My annual fee level of interest would be: $99.00 per year: 1 2 3 4 5 $149.00 per year: 1 2 3 4 5

Name(s)______

Address______ City______ State______ Zip______

Telephone: Home ______ Work ______ Fax ______

We appreciate you giving us this valuable help. You will receive a **10% discount** off your next service order. (Value up to $20.00) A certificate will be mailed to your address as shown above.

Thank you!

Figure 11-2 Great performance questionnaire.

Step 3: Developing a Service Strategy

A service strategy is defined by Fortune Magazine as "knowing exactly which customers you want to serve and figuring out what kind of service will loosen that customer's purse strings." (See Figure 11-3.)

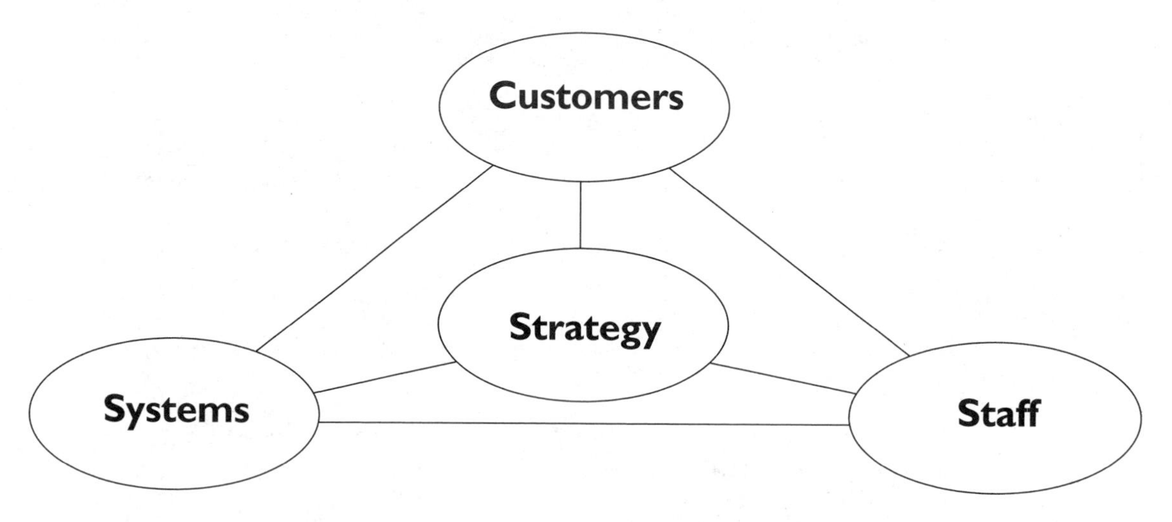

Figure 11-3 Service strategy.

Customers

To articulate that strategy, you must start by describing your target customer in great detail.

As an example, two of our target groups are:

- Single women, head of household, $35,000+ in gross income, General Motors or Toyota vehicle owner...
- Boomers, married with children, 55+ years of age, with combined household incomes of $100,000+...

Identify their wants, needs and expectations in terms of things that are important to them, like:

- Ready when promised
- Fixed right the first time
- Quality
- Convenience
- Price
- Organization
- Cleanliness and order
- Prestige

If you don't know who your customers really are or what they really want, you will never be able to go beyond fixing the broken car, and that isn't enough anymore!

So start with a total quality service-directed strategy at the center of your plan. Then focus on your employees, your systems, and your strategic plan.

Employees

Share your vision with everyone involved in your company. Develop and deliver the education and training necessary to insure that all employees understand just how important delivering consistently high-quality service is and emphasize the role they play in making that happen. Insure that everyone understands that what they do and how they do it will help move the company forward. Make sure everyone realizes that without their total commitment to delivering high-quality service, achieving it is difficult, if not impossible.

Systems

Gather information in terms of the business metrics necessary to help you benchmark performance and facilitate the greatest improvement possible. These metrics are commonly known as Key Performance Indicators (KPIs) and some of them are:

Average Revenue Per Customer	How much each customer spends with you per year
Average Invoice	Average sales per invoice written
Car Count	Number of cars worked on per day/week/month/year
Labor Mix	Ratio of the percentage of labor sales to total sales, compared to percentage of parts sales to total sales
Labor Content Per Job	Average labor content in hours per average invoice
Share of Customer	Your share of the total amount spent on automotive service and repair

S.W.O.T.

Formulate a strategic plan utilizing the S.W.O.T. model for situation analysis. This forces you to look at the *situation*, its inherent *weaknesses*, the *opportunities* it presents, and any *threats* that might accompany any specific action.

Strengths are purchase attributes on which your customers rate you significantly better than the market, and they generally contribute to your competitive advantage—whatever that might be.

Opportunities are purchase attributes on which your customers rate you significantly better than your prospective clients rate their primary suppliers of the same or similar products and services.

Weaknesses and *threats* are the opposite of strengths and opportunities. They constitute areas that need attention and reevaluation. This will, or at least should, result in an ongoing improvement process.

It will take a different kind of process than you may be used to, but establishing this kind of a process will, or at least should, be well worth the effort (see Figure 11-4).

It works something like this. The goals, objectives, and standards of the company power the Customer Loyalty Engine. These goals, objectives and standards result in a set of controls consistent with management's values and guiding principles that can help drive improvement efforts forward without compromising the integrity of the company. The development of the *new*

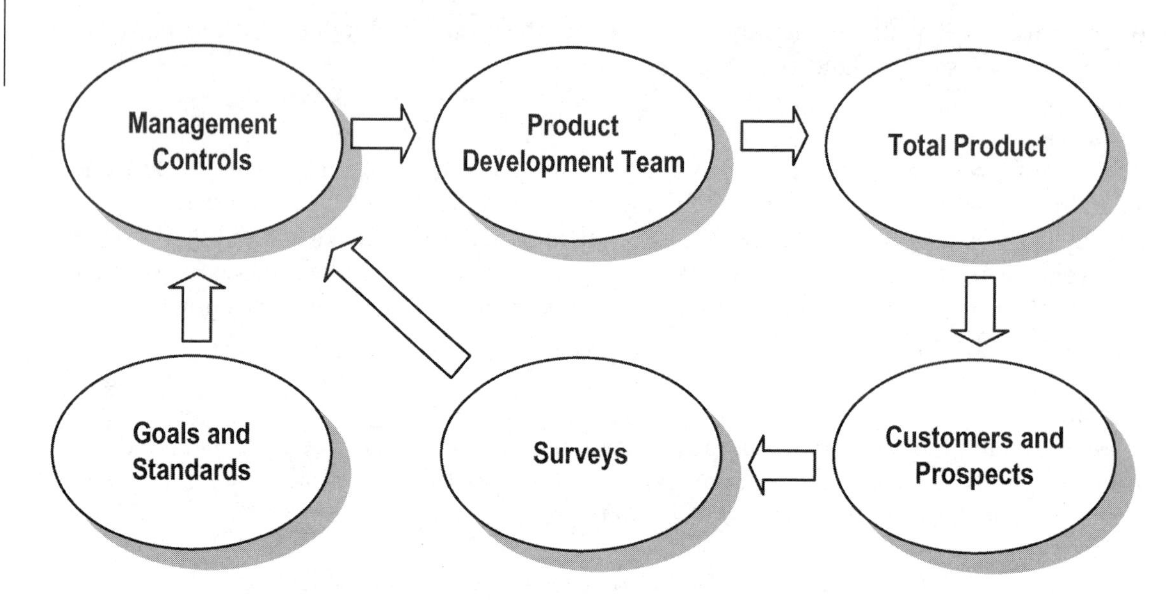

Figure 11-4 The customer loyalty engine.

products comes next, which in our case translates to loosely into different ways to deliver automotive parts and services. This total package of value-added products and services is then offered to select group for evaluation and review which reinforces the company's continuous improvement efforts.

Step 4: Developing a Service Structure

Get everyone in your organization to participate. Being *partially* committed is almost like having *some* integrity, and it just is not enough! In order to insure everyone's voluntary and active participation, allow everyone to participate in the development of this structure and then discuss the importance of a total commitment to the strategy or strategies.

Establish specific objectives and allocate resources (cash!) as necessary. But remember, objectives are specific—they deal with how service is actually delivered. Things like answering the telephone by the third ring, having the vehicle ready when promised, or fixing it right the first time. They are centered on specific actions like free vehicle pick-up and delivery, when the shop opens and closes, or whether or not it is open on the weekends.

To be successful, everyone must know what to do, how to do it properly with speed and efficiency, and why doing it (whatever it is) is important, as seen in Figure 11-5.

Then, all you have to do is create an environment that makes doing it easy!

Operational Excellence = Consistency

My mother's father became very successful in the delicatessen business in New York. Before he died, I had the opportunity to talk to him about his success. Although very young (16 years old, to be exact), I wasn't too young to both understand and remember what he said. He told me that most people couldn't tell the difference between a good corned beef sandwich and a bad one. However, they could see the difference if the sandwich was thick one time and skinny the

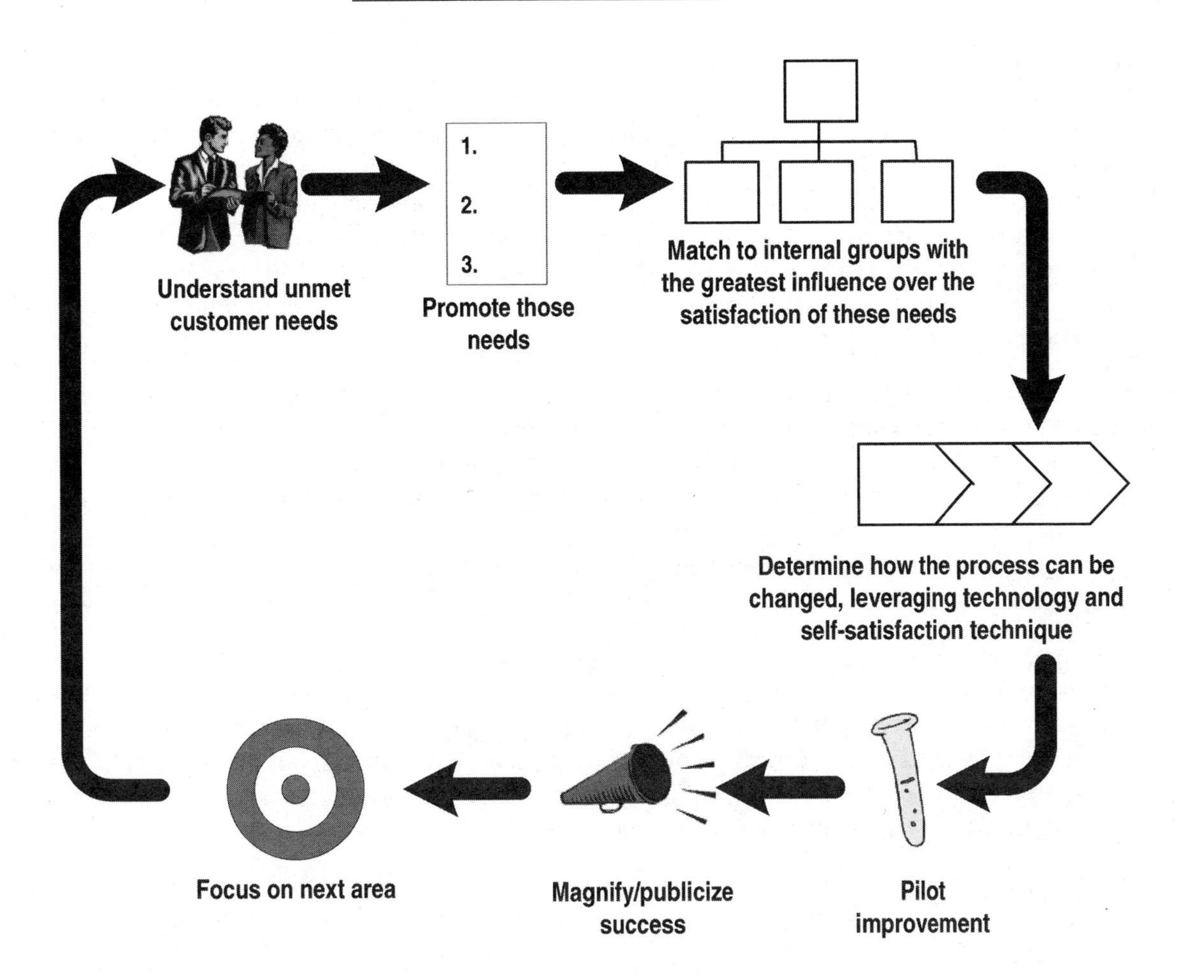

Figure 11-5 The re-engagement process.

next and were smart enough to tell the difference if the sandwich tasted one way one time and a different way the next.

He said that consistency was the most critical ingredient in the formula for success, because without it, replicating what you are doing right is impossible. So, it's far more important to be consistent than it is to be good, because there are too many people who can't appreciate the difference between good corned beef and bad!

The Chinese Definition of Insanity

I shared this with you earlier, so you have read it before: If you always do what you always did, you will always get what you always got!

That's just an updated version of the Chinese definition of insanity, which is doing the same thing the same way every time you do it and somehow expecting the results of your efforts to be different. If that is an accurate definition of insanity, the converse must also be true. That is, insanity can also be defined by doing the same thing a different way every time and somehow expecting the results to be the same!

Consequently, the need for a clear and comprehensive set of polices and procedures is critical.

How would they apply to the automotive environment? Well, we've already agreed that the primary concern of most vehicle owners today is having their car or truck completed when promised, yet I have rarely been in a shop that indicates the time the customer expects his or her vehicle to be completed somewhere on the technician's copy of the repair order. If it was noted, the person working on the vehicle could alert the shop manager or owner if something occurred to prevent the vehicle from being ready when expected. In fact, I have rarely heard shop owners ask the customer when he would like the vehicle completed while the estimate was being written. It is not realistic to expect to have every vehicle in the shop completed on time when no one knows exactly when *on time* is!

Consequently, if *on time* performance is important—and we know that it is—our policy should be to ask the customer when they would like or, at least when they *need* the vehicle completed, and then indicate that somewhere on the repair order. The policy should also include provisions for what to do if or when there is a delay that might impact that delivery time, such as calling the customer to let them know what happened, why it happened, and how it will impact delivery. Perhaps more importantly, that policy should help manage customer expectations by controlling those expectations—such as indicating to customers that appointments are generally made for the whole day, with the earliest arrivals being those vehicles most likely to be completed on time.

We all know that just because the actual work took 3.5 hours, that does not mean the 3.5 hours of work were consecutive. Jobs start and stop and then start again for hundreds of reasons, which basically means that those 3.5 hours of work could have taken six or eight hours to actually complete. The problem is that while we all know that, we do not always act like we know it or communicate that knowledge to the customer!

Another example of how *process* or *systems* impact the pursuit of success is the role both play in fixing the vehicle properly on the first visit, which also relates to the issue of consistency.

A clear example of this can be found in two of the most common and elementary service operations performed thousands of times a day in shops all across the country. One has to do with inspecting brakes, and the other with testing a vehicle's electrical system.

Ask the majority of shop owners across the country how their technicians communicate brake wear or condition to them and you will almost always hear that remaining brake friction material is reported in percentages— "these brake pads or shoes have 10 or 20 percent of the friction remaining." That was the way we did it until we were asked about the accuracy of the percentage by someone pointing out our inconsistencies. In other words, if you don't know how much material is on the pad or shoe when it's new (and it can vary), then how can you know what is left?

I began to think about all the times previous repair orders suggested a vehicle had 10 percent of the friction remaining, only to have 15 percent of the friction reported remaining 5,000 miles later! Did the *friction fairy* come in the middle of the night and put some brake material back? Of course not!

It is more accurate to report the friction material remaining in thirty-seconds of an inch (or millimeters if you prefer) because a thirty-second of an inch is a thirty-second of an inch for everyone. Developing a form such as the one seen in Figure 11-6 will *force* your technicians to give you the information you both want and need to communicate effectively with your customer base, as well as build the kind of consistency that guarantees world-class service. It is the only sure way I know to translate technical skills into a customer-relations asset.

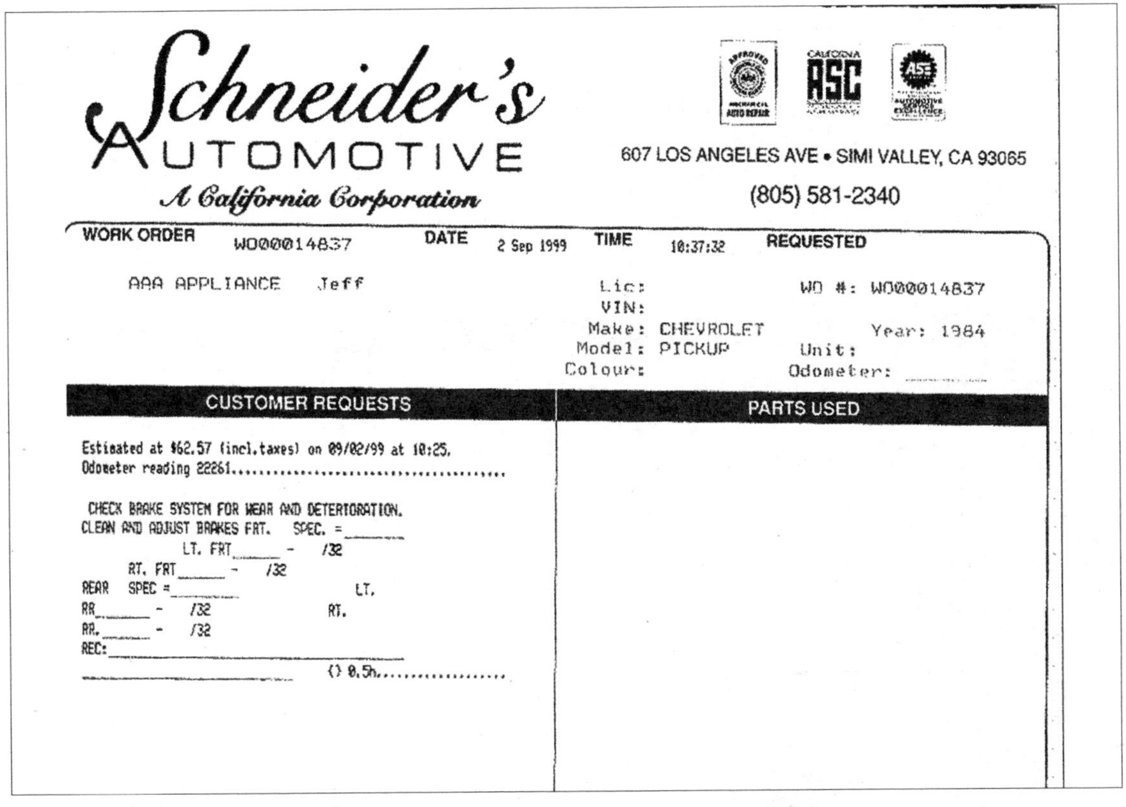

Schneider's
AUTOMOTIVE
A California Corporation

607 LOS ANGELES AVE • SIMI VALLEY, CA 93065
(805) 581-2340

WORK ORDER W000014837 DATE 2 Sep 1999 TIME 10:37:32 REQUESTED

AAA APPLIANCE Jeff

Lic:
VIN:
Make: CHEVROLET
Model: PICKUP
Colour:

WO #: W000014837
Year: 1984
Unit:
Odometer: ______

CUSTOMER REQUESTS | PARTS USED

Estimated at $62.57 (incl.taxes) on 09/02/99 at 10:25.
Odometer reading 22261....................................

CHECK BRAKE SYSTEM FOR WEAR AND DETERIORATION.
CLEAN AND ADJUST BRAKES FRT. SPEC. =______
LT. FRT______ - /32
RT. FRT______ - /32
REAR SPEC =______ LT.
RR______ - /32 RT.
RR.______ - /32
REC:______
______ () 0.5h..................

Figure 11-6 Brake inspection template.

Another example of process and systems applied to service operations is an AVR, or Complete Electrical Systems Test. It sounds comprehensive, but what exactly is a *Complete Electrical Systems Test?*

I've been in hundreds, perhaps thousands, of shops in my lifetime, and I have seen countless technicians check countless electrical systems with the following results scribbled on the technician's copy of the repair order: *bad alt* or *dead batt.*

Is that all the information you need to communicate with the owner of that vehicle? How many times has a rebuilt alternator been installed on a vehicle with a marginal battery, only to have it towed back to the shop six, eight, or ten weeks later with what, to the customer, looks like an identical problem? Doesn't it make more sense to check the *whole* system out completely and then document the total results on the repair order so that you have the security of knowing exactly what shape that system was in when it was tested, rather than having to wonder about what went wrong? Doesn't it make more sense to pose the question of a marginal battery to the owner at the time the vehicle is tested, and let them decide whether or to take the chance? At least that way, if the vehicle comes back, the responsibility lies with the owner for not making the right decision instead of with you—as the service provider—for not providing the right information. Although this makes sense, I have rarely seen a Complete Electrical Systems Test performed the same way twice in the same shop!

The key to solving this inconsistency is a form that *demands* certain key pieces of information. The key is a process that requires each technician to move through the test the same way each time it is performed. After all, that's the only way to create the kind of consistency we are talking about, isn't it?

With a form such as the one seen in Figure 11-7, the technician is guided through a series of steps that delivers the information necessary for both an accurate diagnosis and a complete and accurate estimate to the service counter. While you might decide to change the content of the form itself, I am sure that consistent performance without such a form is difficult, if not impossible!

CUSTOMER REQUESTS	PARTS USED
Estimated at $62.57 (incl.taxes) on 09/02/99 at 10:25. Odometer reading 22261.................................. COMPLETE ANALYSIS OF THE VEHICLE'S 12 VOLT ELECTRICAL SYSTEM TO DETERMINE THE CONDITION OF THE BATTERY/CABLES, ______ STARTING & CHARGING SYSTEMS: CLEAN CABLES AND TERMINALS, CHARGE BATTERY. CHECK INITIAL BATTERY VOLTAGE, ______ CHARGING VOLTAGE ______ LOAD TEST BATTERY AT RATED LOAD FOR 15 SECONDS: VOLTAGE DROPS TO ______ / ______ V @ ______ / ______ AMPS. [GOOD] [MARGINAL] [BAD] STARTING SYSTEM TEST: CRANKING VOLTAGE IS ______ / ______ V AT ______ / ______ AMPS CURRENT FLOW. [GOOD] [MARGINAL] [BAD] CHARGING SYSTEM TEST: NO LOAD VOLTAGE ______ / ______ V. MAXIMUM OUTPUT: ______ / ______ AMPS WHILE MAINTAINING ______ / ______ V. [GOOD] [MARGINAL] [BAD]	

Figure 11-7 Electrical system test template.

Step 5: Find Good People and Treat Them Like People!

Someone much smarter than I am once said that you can't polish mud! You can't force a genuine and sincere smile or a gracious and accommodating attitude! In fact, perhaps the best human resources advice anyone could ever give you is to "Hire for *attitude*. Train for *ability*."

Every business is a *people* business—especially a business like ours that has just moved from being transaction-based to relationship-based. Employees who think the kinds of issues we have been talking about here are foolish and a waste of time or money may not be trainable! Therefore, it would not make any sense for you to spend your time and waste your energy wrestling with a pig. You're going to get dirty and the pig will probably like it!

Frightening Statistics

If finding good people is hard, keeping the people you have motivated and working at their fullest potential is harder as evidenced by the following statistics:

1. Less than one out of four workers say they are currently working at full potential.
2. A full one-half say they work no harder than necessary to keep their jobs.

3. Seventy-five percent feel they could be more effective than they are.
4. Six out of ten feel they are not working as hard as they used to.

It is our responsibility to insure that every employee both knows and understands the important role they play in the continued success of the company.

Who is to blame— them? Or, us?

> *There are still leaders and firms that view fear, distrust, and meanness as desirable management techniques.*
>
> *Managing through fear and building organizations that are filled with distrust is not only inhumane, it is bad business.*
>
> *—Jeffrey Pfeffer & Robert Sutton, authorities on organizational behavior at Stanford University*

Perception Is Reality
Or is it?

- Only 47 percent of employees felt they could challenge accepted practices
- Only 46 percent said that they were sure of their jobs even if they continued to perform well
- Only 45 percent thought management was interested in their well-being
- Only 37 percent said that innovative ideas can fail without penalty to the originating person or work unit.

The Customer Comes Second...
In perhaps one of the best books I've ever read about the importance of surrounding yourself with quality people, the author (Hal Rosenbluth, president of Rosenbluth Travel, one of the largest and most successful corporate travel agencies) suggests that the key to success in any organization is finding the right people and then treating them like people. If you treat employees fairly and give them the opportunity to realize their dreams and achieve their goals and objectives, then they will do the rest! All you have to do is keep them motivated and treat them the way you would want to be treated—as well or better than you treat your customers—and they will take care of the rest.

If you don't, they aren't likely to treat your customers any better than you treat them!

Step 6: Motivate & Train Every Employee

Success comes when every employee believes that high-quality service is the only road to lasting success! The problem is that few, if any of us, often experience quality service. If you don't believe me, just take a Saturday or Sunday afternoon off, head for the local mall, and spend the day looking for quality service in any of the department stores you find there. Odds are you won't find it!

Examples of world class-service delivery are rare, if they can be found at all. Therefore, benchmarking or modeling your service delivery against that of some other company is difficult. If employees practice what they see and experience when they are on the receiving end of service delivery, they are likely to be just as rude and neglectful as the person or persons they were dealing with when they are asked to deliver service that is helpful and considerate.

We should all remain aware of the fact that service professionals are not born—they are created. Service education, training, and motivation must be an on-going process, supported and modeled by everyone in the organization. A pleasant expression or a false smile is no longer enough to satisfy a customer base that has grown unwilling to tolerate *bad* service.

Most businesses don't get it! Businesses remain unwilling to train front-line employees when it is their performance that is critical to the customer's impression of the quality of service. In over 200 businesses surveyed, less than one day of service training per year was offered to people in service-related positions. Training, if available at all, was technical in nature. That model is no longer capable of creating or sustaining the kind of relationships we will need to remain in business—let alone become successful! To be truly service-driven, you must train everyone in your organization—full-time as well as part-time, and long-term employees as well as new hires.

Owners and managers are not the only individuals in the automotive service environment responsible for delivering the kind of service quality we have been discussing. Delivery drivers and technicians are likely to interact with the consumer during the normal course of delivering that service. We should not lose sight of the fact that *anyone* sent to interact with the customer in the normal course of doing business represents your company to that customer.

Consequently, we must choose our vendors carefully! The tow truck driver sent to rescue a stranded motorist, who is in a very vulnerable and helpless state, is no different in the mind of the customer than any one of your uniformed employees and should be chosen accordingly!

Who Does the Training in Your Company?

Who is responsible for training new hires in your company? Is it you, or is it one of your employees? What is, or what will be, your procedure for bringing a new hire into your business? If you are like the thousands of other automotive service providers out there, the orientation process is something like this:

> *"Glad to have you with us! Here is a work order. Put your tool box over there and get to work! I will be out in a couple of hours to see how things are going... If you have any problems or questions, give me a shout!"*

You must ask yourself how much time are you actually willing, or able, to spend with new employees. If you can't spend a great deal of time integrating a new hire into your business, do you have the tools necessary to compensate—such as an employee handbook? Is there a policies and procedures manual available to all new employees, or are they supposed to figure things out on their own?

If you don't take responsibility for training or educating a new employee, someone else in your company *will* accept that responsibility—possibly the employee you would least want to see influence someone new.

Crises... or... Dangerous Opportunities

The Chinese ideogram for the concept of **crisis** (Weijis) is really composed of two characters, as seen in Figure 11-9. As one might expect, one denotes *danger*. However, the other suggests *opportunity*. A crisis could then be considered a *dangerous opportunity*. Whether the opportunity outweighs the danger, or vice versa, is really up to the individual confronting the crisis. This is just another way of saying *perception is reality*.

If you find yourself confronting a customer relations crisis at the service counter, you can ultimately be the one to determine whether the danger will prevail, in which case you lose the opportunity to salvage the relationship. Or you can be the one to take advantage of the opportunity, relegating the inherent danger to a minor position. It's all a matter of control—control of the situation, and control of the self.

Step 7: Empowerment

Figure 11-9 Weiji: crisis... a dangerous opportunity.

We know from all the customer satisfaction surveys done so far that the faster a problem is solved to the ultimate satisfaction of the consumer, the more likely you are to salvage that relationship. I have told countless customers over the years that our relationship—the relationship we share with the vehicle owner—will never be fully defined unless (or until) there is a problem. It's easy to maintain a positive relationship when everything is going smoothly. Skill is required to manage that relationship if and when things start to fall apart!

cri·sis (krī'-sis) *noun, plural* cri·ses (-sēz) **1. a.** A crucial or decisive point or situation; a turning point. **b.** An unstable condition, as in political, social, or economic affairs, involving an impending abrupt or decisive change. **2.** A sudden change in the course of a disease or fever, toward either improvement or deterioration. **3.** An emotionally stressful event or a traumatic change in a person's life. **4.** A point in a story or drama when a conflict reaches its highest tension and must be resolved. *noun, attributive.* Often used to modify another noun: *crisis intervention; crisis planning.*

That means, in order to insure a positive and effective impact on your relationship with a consumer, a problem/argument/misunderstanding must be resolved by the consumer's first contact with an employee. The longer it takes to resolve a problem or satisfy a dispute, and the further up the ladder the employee has to go for an answer, the less effective the result.

That means you must empower your employees to act quickly when you are not there to handle things yourself. There aren't many terms in the English language that can provoke the kind of abject terror a word like *empower* can, but empowerment doesn't have to be that frightening—for you *or* your staff.

Empowerment can be defined as *responsible freedom*—the freedom to act within pre-established guidelines. The key to dispute resolution in the automotive service environment is *responsible freedom.* It's the only way to insure that a problem can be resolved quickly, completely, and (hopefully) to the customer's ultimate satisfaction. To do that, you must empower your employees to make on-the-spot decisions within those pre-established guidelines. This, of course, means that you have to create your own pre-established guidelines!

Benefits of Empowerment...

- When employees are empowered, they are likely to be more enthusiastic about their job.
- They know they have a responsibility to identify problems and they know they have the authority to fix whatever it is that is broken.

To encourage the kind of open and honest relationship between employees and management that results in the kind of teamwork required for success, management must:

- Encourage open communication.
- Praise, pay, and promote people who deliver *bad news* to promote continuous improvement.
- Treat *failure to act* as the only true failure by punishing the failure to take action, rather than punishing actions that are not 100% successful.
- Encourage employees to learn from their failures by sharing those failures with management and with each other.
- Give second and third chances.
- Eliminate people who downgrade, intimidate, or humiliate others.
- Learn from and even celebrate mistakes, particularly when it comes to trying something new.

Step 8: Obtaining the Right Tools and Resources

It is important to realize that poor service and/or dissatisfaction doesn't always originate with the person handling the service interaction with the customer. The blame might belong elsewhere. One of the most common causes of poor service delivery is a failure to provide the necessary tools required to deliver high-quality service.

The Three P's: Policies, Procedures, and Physical Plant

Too often, customer dissatisfaction is the result of inadequate internal systems or the kinds of inconsistencies that result when there is an absence of written policies and procedures... or the guidelines required for consistent service delivery. Another reason for poor service quality could be reliance on outdated or antiquated equipment—the kinds of inadequate physical resources caused by a failure to reinvest in the business. It could also come from inadequate technical training or education— a failure to prepare for a dynamic and unpredictable future.

Failure to deliver adequate service quality could result from a lack of leadership or commitment, or it could be the result of unrealistic goals or objectives. In fact, there is probably no more certain way to guarantee poor or inadequate service delivery than by imposing unrealistic or unreasonable production goals. Is 130 percent shop productivity a reasonable benchmark? Does it help or hurt the facility reach its long-term goals and objectives? If not, what does? Most shops become profitable when productivity exceeds 70 percent. Who benefits when the consumer suffers—especially when the reason for the suffering is inefficiency and lack of organization?

Fortunately (or unfortunately), that is a question you will have to wrestle with on your own.

Step 9: Exceptional Service

To deliver the kind of service quality we have been discussing, you *have* to go beyond the wants, needs, and expectations of your clients and make sure they notice it. To move from customer satisfaction to *customer delight*, you must truly do the unexpected!

You must make the commitment to deliver Total Quality Service.

Step 10: Renew Your Commitment to Service Satisfaction

In his book *Mastery*, George Leonard talks at great length about the phenomenon of *plateau-ing*. In it, Leonard suggests that while mastering something:

> *"You'll probably end up learning as much about yourself as about the skill you're pursuing. And although you'll often be surprised at what and how you learn, your progress towards mastery will almost always take on a characteristic rhythm that looks something like this (see Figure 11-10):*

Figure 11-10 The mastery curve.

> *There's really no way around it. Learning any new skill involves relatively brief spurts of progress, each of which is followed by a slight decline to a plateau somewhat higher in most cases than that which preceded it. The curve above is necessarily idealized. In the actual learning experience, progress is less regular; the upward spurts vary; the plateaus have their own dips and rises along the way. But the general progression is almost always the same. To take the master's journey, you have to practice diligently, striving to hone your skills, to attain new levels of competence. But while doing so – and this is the inexorable fact of the journey – you also have to be willing to spend most of your time on a plateau, to keep practicing even when you seem to be getting nowhere."*

As suggested above, no matter how hard you try to maintain motivation, awareness, or excitement, your service programs will ultimately loose their momentum at some point. Service quality will increase dramatically at first, but will surely reach a plateau. At that point, you must review good customer-service habits and reinforce them with public recognition—remembering that praise, appreciation, and recognition are among the most powerful motivators when it comes to developing human resources.

CHAPTER SUMMARY

Great performance is achieved by understanding what obstacles stand in the way, who owns the obstacles, and how to act as a coach to remove the obstacles. The 10 steps for achieving customer satisfaction are:

1. Making the commitment
2. Knowing your customer
3. Developing a service strategy
4. Developing a service structure
5. Finding good people and treating them like people
6. Motivating and training every employee
7. Empowering your staff
8. Obtaining the right tools and resources
9. Providing exceptional service
10. Renewing your commitment to service satisfaction

How will your company achieve customer satisfaction? Write your thoughts in the chart on the next page.

Thoughts

Actions

Results

CHAPTER 12

Achieving Customer Satisfaction

INTRODUCTION

This chapter introduces you to:

- How to achieve customer satisfaction through a five step process
- A philosophy of continuous improvement (known to the Japanese as Kaizen) that you should embrace in your pursuit of high levels of customer satisfaction

CONTINUOUS IMPROVEMENT

The Japanese word "Kaizen" can be loosely defined as *continuous improvement including everyone*. To achieve high levels of customer satisfaction, you must embrace a philosophy of continuous improvement (Kaizen) in everything you do. You have to remember, however, that customer satisfaction is the result of achieving excellence in every aspect of your business, not just in the area of customer relations.

As an owner or manager, you must create, and then support, quality service and continuous improvement efforts with your own personal commitment—commitment that manifests itself in the education and training initiatives necessary to make these ideals a reality. Then you must develop and enforce a standard of excellence that ensures high quality-service delivery to the loyal and lifetime customer base that will certainly result from these efforts.

It's a Virtuous Cycle

Customer satisfaction is a moving target. Customer wants, needs, and expectations are continually changing. In order to meet or exceed those wants, needs, and expectations you must embark upon a journey that has, as its ultimate destination, the delight of each and every one of your customers every time they are in your care and custody. This process starts with a total commitment to continuous improvement, meticulous attention to every detail of the service delivery process constantly modified and refined—but it doesn't end there. Continuous improvement efforts, in order to be successful, must be reinforced continually. If you are successful, the result is customer loyalty, and with customer loyalty comes business success.

- Continuous improvement
- Continuous reinforcement
- A coupling of continuous improvement and continuous reinforcement
- Customer loyalty—result of continuous improvement and reinforcement
- Business success—direct result of customer loyalty

Remember... quality is the consistent adherence to a customer's expectations. It is achieved by anticipating problems before they develop and creating the infrastructure necessary to eliminate or at least avoid them. It demands meticulous attention to detail, as well as mutual trust and open channels of communication.

This is a process, not a program. A *program* has a beginning, middle, and an end. It starts and finishes. A *process* is ongoing, and this particular process requires the commitment, leadership, and involvement of everyone concerned, as seen in Figure 12-1.

Keep in mind that it takes seven to twenty-two times more of your dollars to attract a new customer than it does to keep an old one depending on just how effective those marketing efforts are! Also keep in mind that it doesn't take much to *provoke* a customer. In fact, more than half of all customers surveyed indicated they *would try* another service provider if they became dissatisfied or encountered a problem not promptly resolved by their current service provider.

It Really Doesn't Take Much

Become dissatisfied with the quality of service you provide your customers now, regardless of how good you think it already is! Continually strive to become the best at what you do. Don't settle for anything less than *exceptional*—exceptional quality of workmanship and professionalism, and exceptional service quality and service delivery.

When You Try a Little Harder...

Work a little smarter... know your customers a little better... and succeed in satisfying them just a little more than your competitors are capable of doing, and you will achieve customer satisfaction. It won't be by default—it will be *by design*!

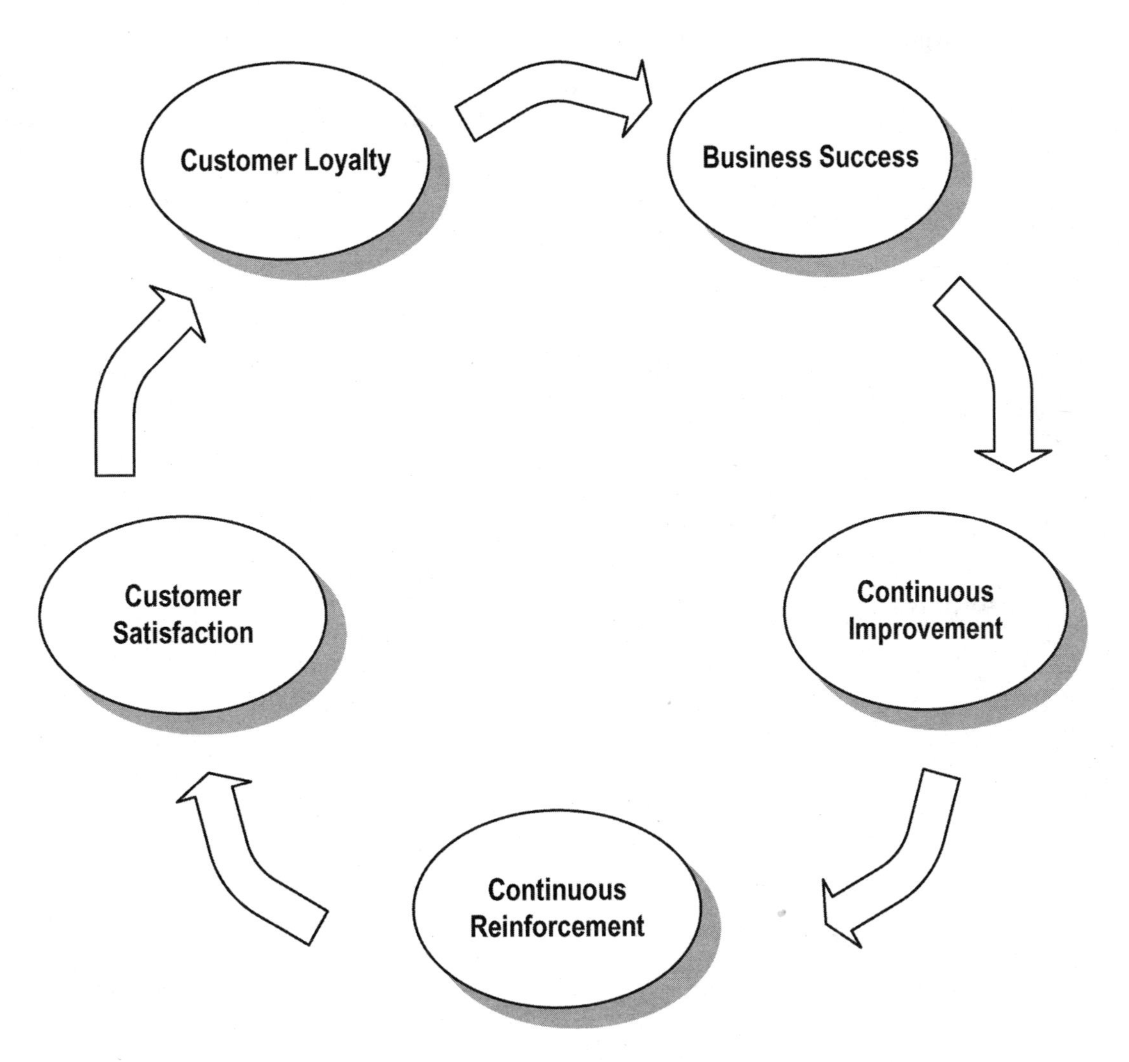

Figure 12-1 The virtuous cycle.

CHAPTER SUMMARY

To achieve high levels of customer satisfaction, you must embrace a philosophy of continuous improvement. Customer satisfaction can be achieved in the five steps illustrated in Figure 12-1: improvement, reinforcement, customer satisfaction, customer loyalty, and business success.

How will your company achieve business success? Write your thoughts in the chart on the next page.

Thoughts

Actions

Results

CHAPTER 13

Customer Loyalty: Is Satisfaction Enough?

INTRODUCTION

This chapter introduces you to:

- The myth behind customers-for-life and understanding that customers have limited loyalty
- Methods for increasing customer loyalty

WHY HAVE LOYAL AND LIFETIME CUSTOMERS?

Profit in business comes from repeat customers... that boast about your product and service.... Profit in a transaction with a customer that comes back voluntarily may be 10 times the profit realized from a customer that responds to advertising...

Dr. W. Edwards Deming
Out of the Crisis

Industry surveys suggest that most consumers of automotive products and service are satisfied 85 to 95 percent of the time. Unfortunately, that does not stop 60 percent of them from patronizing a different service provider (repurchase rates are only 40 percent) the next time they require automotive service.

The goal in any business—but especially in ours—is to both *find* and *keep* customers. That means converting satisfied customers into loyal and lifetime customers. Why? Because the revenue stream generated by a loyal and lifetime customer trading with you for four or more years is generally 3.5 times greater than it is from a new customer!

THE MYTH

The concept of *customers for life* may be a myth. After all, how many customers can you claim are really yours *for life*. In reality, they are only yours for a fixed period of time—customer loyalty is never quite 100 percent. If you do not believe me, look under the vehicles of those customers you consider most loyal and see if everything there is *yours*.

In reality, customers today are:

- Better educated
- More selective
- Have access to more and better information than at any time in prior history
- Likely to move, marry, die, decide they no longer *like* you, and experiment

A more realistic goal would be to capture as much of your target customer's business for as long as you can hang on to it.

LIMITED LOYALTY

We all have to deal with the concept of *limited loyalty*. That means capturing as much of the customer's attention for as long as is realistically possible. If we create realistic expectations and attainable goals, and then dedicate ourselves to delivering the highest-quality products and service available anywhere, we cannot help but succeed!

INCREASING CUSTOMER LOYALTY

Here are a few ways that you can increase loyalty throughout your customer base. Each will improve your prospects for keeping customers longer. When looked at individually, they are not impressive nor are they particularly effective. However, when they are combined, they can—and will—make a significant impact.

Guarantees

A guarantee is perhaps the only tangible way you, as a service provider, have to establish confidence in your products and services. If the customer is searching for security and confidence, the guarantee is your promise that her expectations will surely be at least met, if not exceeded. Consequently, a *compelling* warranty (in terms of time and mileage) is a critical ingredient when it comes to increasing customer loyalty. Remember that the guarantee must be realistic. If it is too attractive, it may lose its credibility. Recognize that a warranty's desirability is defined by the customer, not by the service provider. In that regard, it should have as few caveats as possible. The only purpose of a guarantee or warranty is to establish *trust* in the mind of the purchaser that the *promise* of future performance will be realized and that if it is not, they are protected.

Customer-Convenient Hours

In a world in which time is quickly becoming more precious than money, being open when the customer wants to do business can be critical to your success. Being in a *high-demand* industry can mitigate the need to be open for extended hours—but not forever, and not in all cases. The trade-off between accessibility and *normal* hours of operation must be continually weighed in terms of the demands of the marketplace versus quality of life for the service provider and his or her staff.

Automate the Telephone

Someone once said that if something was invented or developed after you were born, it is technology. If it was invented before you were born, it is a *tool*. The telephone certainly fits those criteria with one interesting exception. As a tool, it has been continually evolving until it has become the thread that ties us together at home, in the office, and on the road.

If we are serious in our customer service efforts, we must respond to our customer's needs when it is convenient for them, not just when it is convenient for us. In most cases, that will mean—at the very least—providing an automated telephone attendant to respond to customers when you are not open.

If you aren't currently providing that kind of a service, try to think of a completed call as a *finished* act—something the customer no longer has to remember. By allowing the customer to take one more item *off their plate*, you provide a valuable, yet relatively inexpensive, service.

Be Accessible

One of the biggest problems most vehicle owners have with the automotive service industry is the sense of indifference they encounter, particularly when there is a problem. If we want to grow customer loyalty, we must be as accessible to our customers as is reasonable, with reasonable being defined by each of us. Being accessible doesn't necessarily mean being there physically. You can publish e-mail, beeper numbers, and possibly home voicemail phone numbers.

Knowing you can be reached in an emergency can result in high customer confidence.

There is one caveat I feel compelled to share with you. Do NOT do this if you don't want to hear the phone ringing at three in the morning!!!! To paraphrase a line from *Field of Dreams*, "If you publish it, they will call!"

FAQs

Every business has Frequently Asked Questions (FAQs), and ours is no exception. Save yourself some *face-time* by publishing these in a flyer, newsletter, or on your website all of which will be discussed in greater detail in *The High Performance Shop*, another book in this series. Offer helpful suggestions based on the questions you are asked most often, and then publish them or incorporate them into your *on-hold* messages. It can result in fewer telephone calls and a customer who feels more confident because she is better informed.

Communicate

I can't stress this enough! Everyone who does what you do is actively and aggressively trying to *steal* your customers. The most effective weapon you have to guard against their incursions is the ability to communicate clearly and regularly.

Stay in touch with your customers; it lets them know that you are thinking about them and appreciate their patronage. Send them helpful information, not just advertising. Customers like to know who you are and what you are all about, not just about what you are selling. The goal is to bring your customers closer to you, your employees, and your business.

Say "Thank You"

In my opinion, you can't say "Thank You" often enough, especially in a society in which loyalty is too often taken for granted. Think about it... when was the last time you received a thank-you note, call, or letter from anyone, for anything? If you did, how did it make you feel? Pretty good, I'll bet. That feeling is no different for your customers. Take every opportunity you can to send a thank -you card or letter. It can result in a customer that feels both recognized and appreciated, two of the most powerful and necessary feelings any of us can experience. And, it is perhaps the least costly, most effective, use of your marketing dollars!

Be Enthusiastic

The only conclusion that can be derived from an encounter with someone who is enthusiastic is enjoyment and satisfaction. No one wants to be confronted with someone who is miserable, mean-spirited, or misanthropic, especially when they have problems of their own (particularly car or truck problems). If your employees are genuinely happy and seem like they want to be there, your customers will more than likely be comfortable being there as well.

Get Online

Develop a presence on the Web, because it lets your customers get to know you when it is convenient for them. Don't lose sight of the fact that the true utility of the Web is access to information—at least for now. Get an e-mail address—it makes you accessible to the customer at her convenience as well. While there hasn't been an overwhelming move toward on-line appointments, that doesn't mean that shift will never occur. Aside from that, it is the least expensive form of communication available to you at this time.

Customer Appreciation Day

It may sound silly or frivolous, but a Customer Appreciation Day is simply a way of saying thank you in more tangible terms. Just remember that if it truly is to be a Customer Appreciation Day, it should NOT be a thinly veiled attempt to drag customers in to buy something. Hopefully, your customers are smart enough to recognize the difference between a sincere desire to say thank you and a thinly veiled pretence to increase sales volume. If they aren't, you should really think about upgrading your customer base!

Recognition

You should take every opportunity to recognize *loyal and lifetime* customers. We did that years ago by identifying certain key customers as *priority service customers*. That was the best way we could find to say "We care!" They are the only customers at our shop that are offered any discounts, allowances, or coupons of any kind! Remember, no one wants to be ignored, forgotten, or taken for granted, so communicate with customers that have been *missing* for awhile—you just may be able to salvage a damaged or broken relationship.

Under-Promise, Over-Deliver

We said it before and we'll say it again: Make realistic promises!

It's easy to over-promise when competition is tough, but it only makes things worse when you can't deliver. Customers do not want to be confronted with any disparities in the relationships they have with anyone that delivers service. It undermines credibility and destroys trust!

Make every possible effort to deliver your products and services before the deadline and under the estimate. The goal is to look better than your competition, not worse!

Listen

Listening is one of the most valuable skills you can develop. It is all about the passive acquisition of useful information, the ultimate purpose of which is to improve service quality. It is passive because it requires nothing more than paying attention on your part. But, it also means allowing your customers to talk about what is important to them, even when it may have little or nothing to do with selling your service. It also means developing a genuine interest in your customers, their families, and their lives. This is, after all, a *service* business, and service is a contact sport! It is interactive, and that means dealing with other people.

Respond

React quickly to your customer's needs, wants, and expectations on the basis of *their* priorities, not yours. A prompt response means someone cares, so... you can figure out what a delayed response means.

Surprise Your Customers

It's easy to forget that somewhere in all of us lives a little kid who loves to be delighted with a pleasant surprise. Sending a birthday card or a note of congratulations or condolences shows that you care and reinforces the fact that your customers are recognized and appreciated for the relationship they share with you and not just for the impact they have on your bottom line.

KISS

Some people suggest this acronym stands for *Keep It Simple, Stupid*! I have a problem with that, especially if the advice is offered to someone genuinely seeking guidance. Seeking guidance isn't the act of a *stupid* person, it's really the first step on a journey to self-actualization. Consequently, I've changed the acronym to mean *Keep it Simple...and Straightforward*! Stated another way: don't get too cute or too clever in your quality service efforts or in the relationship you share with your customers. Do whatever it takes to make it easy to patronize your company. Take a *Yes, we can...* attitude.

Do you sell tires? If you don't, develop a relationship with a service provider who does. Then, act as the *gatekeeper*. Have your clients drop their cars and trucks off at your facility and you take care of getting the tires mounted and balanced. The same holds true for exhaust system specialty work or air-conditioning sublets. Allow your customers to come to depend upon you for anything automotive-related.

Don't let policy or procedure destroy a relationship. Remember what Peter Drucker said: "The purpose of business is to find and keep customers!" Don't lose a quality relationship because your only response to a problem was *going by the book*. The client isn't interested in policies and procedures—the client is interested in positive results, and positive results are only positive as seen through the eyes of the customer.

Differentiate

A *commodity* is an undifferentiated product or service. Your job as an owner or manager is to make your business look different from other automotive service businesses, as much as you are able. If you fail, your business will be lumped in with every other automotive service business in the neighborhood, with price being the only differentiating factor the vehicle owner can use to judge value. Consequently, it becomes your responsibility to make your business stand out. Make every effort to have your business *look* and *feel* different from the consumer's point of view in everything that you do.

Study your competitors. See the products and the services that they offer and how they make those products and services available to their customers. Then, look closely at what they are unable or unwilling to do. Find ways to perform in the vacuum left by what they are unwilling or unable to do. You may find that it's a host of little things that can and often do make all the difference in the world.

Business as usual is a prescription for extinction in today's business environment! You can't maintain your marketing position when everyone and everything around you is constantly changing!

Be Decisive

Make on-the-spot decisions when on-the-spot decisions are necessary! People are no longer willing to listen to "I'll have to get back to you." Empty statements designed to patronize the customer into accepting inaction as action, such as "let me check," "I'm not sure," or "gee, he's not here right now" just won't work anymore (if they every did).

These phrases are designed to do one thing and one thing only, and that is to deny a consumer what she really wants.

Make it Fun

This can be hard to do when someone is despondent because her car or truck is broken,... but do the best you can to let your customers know that you will do everything you can to make the experience as painless as possible. Keep them informed throughout the process, and try not to make them feel as if their automotive service crisis is as painful and distasteful for you as it is for them!

Be Creative

Introduce as many *new* innovations, products, and services as you can—as often as you can.

Paint the shop, re-stripe the parking lot, buy a new sign, or change your work orders or your uniforms. A new look and a new format can increase sales significantly. If nothing else, it could make you and your facility more visible.

Make it Clear

Make sure that your invoices are written clearly in a language that your customer will understand. Your invoice is the single most important communication you share with your clients. It should serve as a written record of everything that transpired between you. Remember to document and explain everything you did and, if possible, why you did it—it will help justify the cost. Look at the concern-cause-cure model we used earlier in this guide, and try to understand how using it effectively will meet the above criteria while creating *value* in the mind of the customer.

It may be hard to listen to, perhaps even hard to articulate, but you get paid by the word—the description of the product or service should be at least as *big* as the product or service itself.

Personalize Everything

Personalize all communications. Every communication should be addressed specifically to the client receiving the message, and you should do whatever you can to insure that it DOESN'T look like bulk mail. All of us get more than enough junk mail, and just because it comes from you does not make it any more appealing!

If the customer information is not 100 percent accurate and correct, do NOT send it! Sending out sloppy marketing materials translates to sloppy work in the customer's mind.

Create Value

Show your customers how to save money. Don't wait for the customer to ask—take the initiative in this relationship. Become a pro-active customer advocate by actively showing clients how to save. If they understand how their relationship has benefited them long-term—how they have saved something important to them, like time or money—they will be less likely to leave for a lower-stated price because they will know you are watching out for their best interest.

Answer Every Telephone Call

In every business there are some calls that fall through the cracks. You never know what might result from the one call you don't find the time for or make the effort to answer or return. Develop a reputation for being a good *listener*—something few of us are trained to do. Take the time to talk to people. It shows you care.

Provide Added Value

Do something extra—something unexpected—but remember, it doesn't do any good unless you tell the people who benefit from whatever it is you did that you did it! We wash every car we work on and have cases of lollipops and candy on hand for the kids (and some of the parents!) who come into the shop—a friend adds a personalized, handwritten thank-you note and a fresh carnation to every finished vehicle. Be creative; do the unexpected!

Answer the Phone Properly

For years we answered our telephone with: "Schneider's!" as if everyone knew who we were and what we did…

Guess what? Not everyone did!

Now I answer every phone call "Schneider's Automotive Repair... this is Mitch. I can help you?"

Slow down... Take a deep breath before you answer the phone no matter how harried or prevailed upon you feel at the moment. Don't make it seem like you are in a rush to get off the phone, even if you are. Your customer will be able to sense that you're in a hurry. They will be able to sense your frustration or annoyance with being bothered.

Customers (and potential customers) are entitled to your undivided attention and concentration. They pay for it every day! Remember, the telephone is the most vital link between you and your customers—one of the most versatile and useful tools in the customer service toolbox.

Do Something!

Do something, anything—even if it is wrong!

It is often easier to recover from taking the wrong action than it is to recover from taking no action at all. All of the above are proven *techniques* guaranteed to increase customer loyalty—but no tool, no matter how well engineered or designed, will work if you don't use it! In order for any or all of these tools to work, you will have to take them out of the tool box and use them!

If you don't utilize the tools we've shared here, be reasonably certain that one of your competitors will!

Customer loyalty is possibly the single most critical ingredient for success in today's customer-relations focused marketplace. Do whatever is necessary to insure it!

CHAPTER SUMMARY

The concept of customers-for-life is a myth. Customers today are better educated, more selective, have more information available to them, will move on to another business, or have a life change which moves them away from your business.

Every business retains customers for a limited amount of time. You can increase customer loyalty by offering guarantees, having customer-convenient hours, automating the telephone, being accessible, offering FAQs, communicating, saying thank-you, being enthusiastic, getting online, having Customer Appreciation Days, recognizing loyal and lifetime customers, under-promising and over-delivering, listening, responding, surprising your customers, following the policy of KISS, differentiating yourself from other businesses, being decisive, making it fun, being creative, answering every telephone call properly, providing added value, and by doing something!

How will your company increase customer loyalty? Write your thoughts in the chart on the next page.

Thoughts

Actions

Results

CHAPTER 14

Customer Retention

INTRODUCTION

This chapter introduces you to:

- How to retain customers through customer driven marketing
- The importance of customer relationships and loyalty in higher-level marketing
- How human behavior indicates future customer relationships
- How to understand and focus your investments on the highest possible return
- How to increase customer loyalty through communication
- How to understand competitors for your benefit

WHY WE ARE IN BUSINESS

The purpose of *being* in business is to create and accumulate wealth. That's what capitalism is all about. According to Peter Drucker, we do that by *making customers* and creating innovative products and services. I'm not sure, however, that the definition is adequate anymore. I would suggest that the purpose of business is to: *make and keep the right kind of customers by solving their problems at a profit with innovative products and services they both want and need.*

If my definition is correct for our industry (or any other, for that matter!), the focus of almost all marketing efforts has been off the mark for years! Historically, marketers in the automotive service industry have focused almost exclusively on attracting *new* customers. Whether through conquest (taking a customer away from the competition) or prospecting for new customers (searching for that elusive first-time customer), the focus has been on continually increasing market share. Market share represents your sales in relation to the total sales for that segment of the market, or your actual number of customers in relation to the total number of potential customers in your market area.

INCREASING MARKET SHARE

Expanding your share of the market is a principle of marketing that is really hard to fault. After all, five to nine percent of your current customer base will defect or desert you every year! They will decide they don't like you anymore, decide they don't like the way they are being treated, try someone else to find out if they are being treated the way they should be treated, move, marry, or finally head for that *big garage in the sky*! Consequently, it is not only prudent but necessary to focus at least a portion of your marketing budget (pre-supposing you have a marketing budget!) on new customer acquisition. I would be lying to you, however, if I told you that it should be the focus of *all* of your marketing efforts and that *all* of the evidence supports my position.

The key to success in our business is not only bringing people to your door, but getting them to spend money and then return regularly over the lifetime of your relationship. The key is to turn first-time customers into profitable customers and profitable customers into loyal and lifetime customers by managing the full scope of your relationship with each and every customer you choose to keep, both carefully and well.

It doesn't mean keeping all customers—it means keeping the *right* customers. It doesn't mean maintaining the relationship at a loss—it means serving the customer at a profit. It doesn't mean inviting everyone to your door—just those individuals, families, or groups with whom you would like to develop and maintain long-term relationships. It certainly doesn't mean focusing all your efforts on an ever-increasing number of first-time visits—it means expanding the personal and commercial relationship you have with those people who have already established a buying pattern or a historical relationship with you and your company. It doesn't mean focusing exclusively on new customer acquisition—it means harvesting what you have already planted by focusing on *customer retention*!

Every year, businesses like yours and mine lose countless potential loyal and lifetime relationships by not focusing their attention in the right place. They treat existing customers indifferently with unresponsive service and products that fail to meet the customer's expectations. They forget that the largest chunk of their marketing investment is spent on getting that new customer to show up, and in many cases they squander that investment by not making sure that the first visit is followed by a second, third, and fourth. They forget that it takes seven to twenty-two times more of your marketing dollars to make a new customer than it does to keep an established customer and that depends upon just how effective that marketing really is. Finally, they forget how time-intensive starting a new relationship can be.

Profits come from achieving operational excellence in all areas of your business, and marketing is no exception. Expanding an existing marketing relationship with the customer is far more efficient and productive than starting a new relationship with a stranger.

CUSTOMER-DRIVEN MARKETING

Once you recognize the importance of retaining customers, understanding their wants, needs, and expectations becomes infinitely more relevant. It is at this point that your marketing efforts should become customer-driven.

To get customers to come back, you've got to give them a reason. Returning has got to be something they *want* to do rather than *have* to do. If they want to, return they will—despite the fact it is expensive, unbudgeted, inconvenient, or uncomfortable. If returning to your facility is something they feel they *have* to do because they have no choice, they will bolt the minute they recognize that they have another choice available to them!

Consequently, customer retention has to be tactically driven, strategically motivated, and based upon past and present customer behavior.

Relationship, Loyalty, and Database Marketing

Customer retention is built upon relationship marketing—that is, marketing to your existing customer base rather than mass marketing, which is focused on the total universe of potential and existing automotive service customers (many of whom you do not need and would not want as customers!).

Relationship marketing's core activities also include loyalty marketing (which is the creation of *relevant* and *intensely personal* relationships delivered with high-touch, world-class service), permission marketing (which occurs when the customer *responds in some way* to your conventional marketing efforts and agrees to receive e-mail or otherwise participate with you based on your direct marketing efforts), and data base marketing, (which uses your own customer database and/or additional data resources to help you identify, and then better understand, customer wants, needs, expectations, preferences, patterns, etc.).

Relationship marketing also recognizes that if you treat all of your customers *equally*, you treat them all *unequally*, because not all customers are equal! Each of us will attract a different customer base based upon any number of variables, including personality, overall business philosophy, and the definition of good service at each facility. Within each customer base, there will be different groups of customers—each with its own unique set of wants, needs, and expectations. Some of these groups will be more interested in reliability than in performance. Others will be more interested in performance than economy. Just as their wants, needs, and expectations are different, their responses to your marketing efforts will be different as well.

HUMAN BEHAVIOR

Past and current behavior is the best indication (predictor) of future behavior!

The basic premise of customer retention-based marketing suggests that it is more efficient (and thereby more effective) to focus your marketing efforts on your established customer base. A second, and almost equally important, premise suggests that past customer behavior is the best predictor of future customer behavior. In other words, customers who have reacted in a positive way to certain aspects of your marketing efforts are likely to react in a positive way to similar marketing efforts in the future. It also suggests that your customer base is subject to the Pareto Principle (the

80/20 Rule that suggests that 20 percent of your clients are responsible for generating 80 percent of your profits, etc.).

Two Groups of Potential Clients

You have two groups of potential clients. One fits your target demographic for age, education, sex, and income, but has never purchased automotive service from an independent repair shop before. The second group is *clearly* outside of your target demographic, but *has* purchased automotive service from independent repair shops in the past. The question then becomes, if you are an independent repair shop, which group is more likely to respond if it is sent a *20 percent off* coupon from your company?

In other words, where would your marketing dollars be better spent—on the group that matches your target or on a group that doesn't match your target but has responded in the past to other service providers in your service delivery segment? Which group would provide the most dramatic response—the best result?

The answer is simple. *Actual behavior* is a better predictor of future behavior than demographic characteristics. Your money would be better spent on the group that has sought automotive service from the independent service aftermarket in the past because the people in that group are more likely to exhibit that same behavior in the future.

There is a critical point to make here that relates back to the concept of *mining* your own database for information. If you accept the above as true, you may be able to predict when a customer is about to leave you by watching their behavior. Once you can predict behavior, you may be able to mitigate that behavior and retain the customer

Active Customers

Even though the purchase of automotive service is a *grudge* purchase (a purchase few vehicle owners *want* to make), active customers can still be considered *happy* customers. They are, after all, happy with you, your service, the way they are being treated, and the way that your products and services meet or exceed *their* wants, needs, and expectations or they wouldn't be coming back.

Jim Novo, author of *Drilling Down*, suggests that active customers like to *win*. They like to feel in control and smart about the choices that they have made. *Good marketers*, according to Novo, take advantage of this by offering the customer something that helps her feel good about the choices she's made.

The reward for participation can range from discounts on selected products and services, a sweepstakes (a favorite of companies heavily involved in e-commerce or direct mail), to *loyalty* programs such as frequent purchaser programs (like airline frequent-flyer programs) that offer incentives for greater participation.

HIGHER-LEVEL MARKETING

Higher-level marketing approaches that *help* the customer feel better about the purchasing decisions she has made can include *warm and fuzzy* elements, such as birthday cards, anniversary cards, or even anniversary cards recognizing the first time she came to your facility. More conventional marketing pieces could include thank-you notes and newsletters.

Regardless, all of your marketing materials should be focused on one overriding purpose—to get the customer to *do something*, even if that thing is to feel good about returning. To do that, you must first ask the customer for *permission* to engage them in this kind of a dialogue and then reward them for participating. Most importantly, whatever you do must include a compelling *call to action*.

Retaining customers means keeping them *active* and *happy*, recognizing ultimately that if you don't someone else will!

The cold, hard, uncomfortable truth is that everyone will experiment eventually. Your job is to help them want to return after they have—and, of course, to keep your relationship with the customer as *profitable* as you can for as long as possible!

Technology of Customer Retention

The technology of customer retention is all about action, reaction, feedback, and repetition. In the first step (action), you listen to what your customer-based research is trying to tell you about how, when, and why your customers are doing the things they are doing. First, formulate a strategy that results in a certain action on the part of that customer group or segment (something you want that customer to *do*), and then take the appropriate action by asking for the desired result in person, through the mail, or on the Internet.

The second step (reaction) monitors the reaction of that segment of your customer base to see if action is taken and if it is, by whom and to what degree.

After you monitor the reaction of your customers, the third step (feedback) is to analyze the information you now have to help determine what those customers are telling you. Once you have that information in your hands, you can reformulate a strategy and begin the fourth step (repetition), which repeats the process all over again.

Critical

Database marketing is a very high-level and extremely effective strategy when fully understood and implemented, but there are some caveats. For it to be effective, it must be bi-directional—a term all of us should understand after years of having to deal with the computerized electronic engine controls found on virtually every vehicle that has been built for the last twenty years. Bi-directional means information has to flow easily from the customer to the business and from the business back to the customer. It also means that someone at the shop, in the business, has to *listen* to what this information is trying to tell you!

As an example, we will look at an illustration of how customer behavior can *talk to us*. We will cull our customer data for a list of all those customers who have made at least two visits to our shop. We can call the period between the first visit and the second visit *latency* because nothing happened during that period. For our purposes, we can say that for the majority of our customers, that period of latency is approximately six months (180 days). The actual period doesn't really matter. Our customer data has already told us something very important about our relationship with our customers and potential customers—it has told us that if a first-time customer is likely to become a regular or lifetime customer, there will be a second visit within that 180-day period.

Now we can look at first-time customers who have NOT been back within the 6-month period. The data suggests that the second group's behavior is not normal. They have not returned within the same period of time that is *normal* for the first group. From this abnormal behavior, a number of questions become evident.

First, why have those individuals failed to return? Are their wants, needs, and expectations somehow different? If they are, how different are they and what are the differences? Are these customers we would otherwise like to become a part of our regular customer base? If they appear desirable to us, how do we get them to return? Do we send them a letter? Should the letter be accompanied by an *offering*—a discounted product or service? Would sending a newsletter—print or electronic—be just as effective?

DATA INTIMACY

In every case, the data is speaking to you. It is telling you that there has either been an action or inaction on your part or on the part of the consumer. All you have to do is learn to listen, and then figure out what you need to do to get the desired result. Finally, recognize that success in anything requires the dedication of resources, and that as an investment, you should be interested and focused on the highest possible return.

Return On Investment

Most marketing and advertising dollars are poorly allocated and, consequently, result in very poor performance. "On average, the return on investment of advertising is one to four percent. By our measures, this suggests that most companies would be better off taking their ad dollars and putting them into certificates of deposit." The authors Kevin Clancy and Peter Krieg, in their book *Counterintuitive Marketing,* suggest that it isn't the principle of advertising that needs to be overhauled and made more efficient or effective, it is how those campaigns are conceived and then implemented that renders them ineffective.

You can increase efficiency by choosing what you do, how you do it, and who you do it for very carefully—choosing to allocate resources where they get the highest possible return on those customers or marketing activities certain to return a dividend.

Instead of spending the same amount on every customer, spend more on some and less on others. The key is identifying where you will get the highest return.

The Data Is A *Listening* System

You can keep your budget *flat* or decrease it slightly if you continuously monitor your marketing efforts to insure they are focused on the most profitable use of your money and your time and away from activities that do NOT result in the same high return on investment (ROI). Instead of spending the same amount on every customer, spend more on some and less on others.

Where do you get the extra dollars when dollars are in high demand everywhere in your business? You take them away from marketing efforts previously focused on your more marginal customers. You *migrate* and reallocate resources toward higher ROI efforts.

The key is identifying where you will get the highest return on your investment. Loyal and lifetime customers spend more the longer they remain loyal and lifetime customers.

Everyone *wants* first time customers to become loyal and lifetime customers, but is this part of your strategic planning? Most small businesses like ours focus on increasing market share or the overall size of their customer base—not on their share of each customer. These strategies are focused on a particular brand or product and not toward a specific market or customer segment. Their purpose is to *create new* customers, not to *retain existing* customers!

They lack understanding and an integrated strategy focused on customer retention—a quantifiable *feedback* mechanism that can be used as criteria for your success.

INCREASING CUSTOMER LOYALTY

In the beginning, there were neighborhoods in which all businesses were Mom and Pop businesses. It was a time of personal service and a high degree of customer recognition and appreciation, all of which resulted in a high degree of customer loyalty.

Today, we find the world in which we live increasingly distant, impersonal, and unfriendly. It is an age of feeling lost that has all of us feeling invisible. The neighborhood grocery store has been replaced by a supermarket forty times larger with infinitely more choices and virtually no personal contact. With the evaporation of personal contact has come an equal, if not greater, loss of customer loyalty.

To increase customer loyalty, we must practice database or Customer Retention Marketing (CRM). CRM is made possible by increasingly powerful and sophisticated computers and software that help record and monitor customer preferences and personal information in an effort to build greater value and more easily deliver world-class service for the customer.

The goal is increased *stickiness*—or what could just as easily be called *emotional loyalty*. Emotional loyalty results in higher referrals, repeat sales, and higher profits.

The Key… Inexpensive Strategies

Focusing on the customer will only be successful if the customer perceives value in the services you offer and associates that value with being a part of your extended *family* of satisfied customers. If this does not occur, all the coupons, special offers, newsletters, and promotions in the world will mean nothing!

There are a number of inexpensive, yet very effective, strategies that will work. You can implement these individually or in conjunction with one another, and we will discuss these in much greater detail in another volume in this series. For now, we will just take a quick look at what you can do to keep your customers close to you.

You can create and implement a preferred customer program, with special offers, discounts, promotions, and preferential treatment *for members only*. You can actually purchase frequent flyer miles from a number of the major air carriers so that your customers can use the rewards they receive as the result of a relationship with you and your company for travel, recreation, or business–whatever they deem appropriate. You can acknowledge customers who have appointments by placing their names on a welcome board to let them know that you are looking forward to their visit and know that they are coming. You can recognize their visit with a thank-you note or follow-up phone call to insure service satisfaction. You can send out event-oriented communications in a newsletter regarding things the customer is specifically and genuinely interested in. Whatever you do, you will want to utilize all the information you have at your disposal to communicate with clients on a very personal and individual basis.

Remember, two of the most powerful human needs are the need for recognition and appreciation. People want to be recognized and appreciated as individuals, not just as statistics!

Critical!!

You can't know what your customers really want, need, and expect from you unless you ask! Ask your customers what's important to them. Ask them what they want, need, and expect from the relationship that they share with you. Then, offer something of value to them in return for their cooperation. Five questions to ask are:

1. What do you like about doing business here?
2. Why did you come to us in the first place?
3. What problems did you have with other places before you started trading here?
4. How did we solve those problems?
5. How are things better for you now?

The answer to that last question is critical to your success! It is what a *positive result* looks like to a customer. It's going to look the same to your other customers and prospects when you tell them about it.

Remember that when you are unsuccessful in your efforts to make *and* keep the *right kind* of customers by *solving their problems at a profit* with *innovative products and services* they both want *and* need you lose in two ways. First, you lose the revenue stream derived from that relationship, and second, your competitor gets it!

Go to your customers and ask them what doing business with your company has done for them and how they have benefited from the relationship.

Finally, ask yourself and the other important people involved in your business the following:

- What are we doing right now to deserve our customer's business?
- In terms of the buying process, how can we ask them to define their wants, needs, and expectations more clearly?
- How can we use what our competitors are doing to our advantage?

Scout Your Competitors

Scout your competitors to see what they are doing. Try to determine what is working and what might not be working. *Mystery call* the automotive service facilities—*all* of the automotive service facilities—to check out their telephone etiquette. Call and tape *after hours* messages. Have friends or associates call and tape *first-time* customer calls.

Interview other shop owners customers, and potential customers. And when you are all done, start the process all over again, because that is exactly what it is—a process.

Remember that while a process might have a beginning—a discernible origin—it should never have an end.

The process of increasing the quality of service delivery available at your facility to a level of world-class performance, the process of knowing and understanding customer wants, needs, and expectations better than anyone else in your market—or perhaps even in the industry you serve—and then translating that knowledge into high customer loyalty, high customer retention, and high ROI is a journey of continual improvement. It is a journey that requires character and commitment, understanding, dedication, discipline, and determination. It is not necessarily an easy journey, nor is it likely to be a particularly comfortable one. But, it is not, however, without reward.

CHAPTER SUMMARY

The purpose of business is to make and keep the right kind of customers by solving their problems at a profit with innovative products and services they both want and need.

It is necessary to focus a portion of your budget on new customer acquisition. These customers, consisting of both personal and commercial relationships, should have already established a buying pattern and a historical relationship with you and your company.

Customer retention is built upon relationship, loyalty, and database marketing rather than mass marketing. It is more efficient and more effective to focus your marketing efforts on your established customer base first.

How will your company successfully retain customers? Write your thoughts in the chart on the following page:

Thoughts

Actions

Results

CHAPTER 15

Conclusion

There is great opportunity for success in the automotive service industry. It may not be obvious and it may not be easy, but it is there, nevertheless. Apply what you learn. It is bound to make the journey somewhat easier. If nothing else, it will make it more interesting!

Recognize that improvement may be incremental at first, and that any goal worthy of the effort and discomfort that comes with change requires both endurance and patience.

To paraphrase the biblical injunction, the road to business success starts by recognizing how important it is to...

> *"Know your customers as you would know yourself...*
>
> *Only know them better!"*

So, take a deep breath! You have already taken the first step. From here, it's just placing one foot in front of the other!

MORE TO COME

For our purposes here, we have divided automotive shop management into three general areas of concentration. As we move through these general areas in the subsequent guides in this series, we will be looking at hiring practices, productivity, service management, and the cost of doing business—including cost/value ratios, value-added services, and hazardous materials communication. In addition to exploring our relationship with the consumer, we will also look at sales and marketing.

Our goal is to create the most comprehensive resource for automotive shop management available anywhere—a complete and exhaustive look at all the hard and soft skills required to manage any business, as well as the tools required to run an efficient, effective, customer-responsive automotive service facility.

The series was not conceived or formulated in a vacuum. It did not originate in theory first and then find its way into reality. It is the result of a lifetime spent in the automotive industry backed up by four generations of genetic memory. I invite you to take whatever you can from these pages. Take the key performance indicators you find here and create a *snap-shot* of your business at the beginning of this journey. Then, see how integrating what you find within the pages of these guides changes those numbers—how it impacts the bottom-line. Finally, let us know how you are doing!

If you have any suggestions for subsequent subjects or ways we can improve what we have already presented, let me know. My goal is provide you with answers to the questions I had to search for or create myself. My goal is to insure that you don't find yourself struggling the way I had to. My goal is to see that you are richly rewarded for your efforts and your ability.

With that said, read this Automotive Service Management series one guide at a time or all at once. See what you like. See what you aren't all that comfortable with. *Then read it again!* Find those principles that make sense to you. Create a plan to integrate them into your business. *Then read it again!* Make the changes required to take your business to the next level. Monitor your progress. Document everything... *and then read it again!*

If it works for you, let us know. If you think of ways we can help you make it work more effectively, let us know!

The American Heritage Dictionary of the English Language, 3rd ed.

Deming, W. Edwards. *Out of the Crisis*. Boston: Massachusetts Institute of Technology Press, 2000.

Drucker, Peter. *On Creativity, Innovation, and Renewal*. New York: Jossey-Bass, 2002.

—. *On High Performance Organizations*. New York: Jossey-Bass, 2002.

—. *On Leading Change*. New York: Jossey-Bass, 2002.

—. *The Practice of Management*. New York: HarperBusiness, 1993.

Jennings, Jason, and Laurence Haughton. *It's Not the Big That Eat the Small, It's the Fast That Eat the Slow!* New York: HarperCollins, 2001.

Leonard, George. *Mastery: The Keys to Success and Long Term Fulfillment*. New York: Plume, 1992.

McKenna, Regis. *Real Time Preparing for the Age of the Never Satisfied Customer.* Boston: Harvard Business School Press, 1997.

Novo, Jim. *Drilling Down: Turning Customer Data into Profits With a Spreadsheet.* Bangor, Maine: Booklocker, 2001.

Pfeffer, Jeffrey, and Robert I Sutton. *The Knowing-Doing Gap*. Boston: Harvard Business School Publishing, 1999.

Reichold, Fredrick F., and Thomas Teal. *The Loyalty Effect.* Boston: Harvard Business School Press, 1996.

Sewell, Carl. *Customers for Life*. New York: Batnum Doubleday Dell Publishing Group, 1991.

Woodruff, Robert B., and Sarah Gardial. *Know Your Customer.* Boston: Blackwell Publishers, 1996.